AF361636

The 15th International Conference on Hand-Arm Vibration

The 15th International Conference on Hand-Arm Vibration

Editors

Christophe Noël
Jacques Chatillon

MDPI • Basel • Beijing • Wuhan • Barcelona • Belgrade • Manchester • Tokyo • Cluj • Tianjin

Editors

Christophe Noël
Institut national de recherche
et de sécurité (INRS),
Vandœuvre-lès-Nancy
France

Jacques Chatillon
Institut national de recherche
et de sécurité (INRS),
Vandœuvre-lès-Nancy
France

Editorial Office
MDPI
St. Alban-Anlage 66
4052 Basel, Switzerland

This is a reprint of articles from the Proceedings published online in the open access journal *Proceedings* (ISSN 2504-3900) (available at: https://www.mdpi.com/2504-3900/86/1).

For citation purposes, cite each article independently as indicated on the article page online and as indicated below:

LastName, A.A.; LastName, B.B.; LastName, C.C. Article Title. *Journal Name* **Year**, *Volume Number*, Page Range.

ISBN 978-3-0365-7806-4 (Hbk)
ISBN 978-3-0365-7807-1 (PDF)

Cover image courtesy of Miserez Design -Schautterstock

Contents

Preface to "The 15th International Conference on Hand-Arm Vibration"

This publication summarizes the proceedings of the 15th International Conference on Hand-Arm Vibration, held on 6–9 June 2023 in Nancy, France. This multidisciplinary conference was organized by the INRS (National Research and Safety Institute, France) under the patronage of the International Advisory Committee on Hand-Arm Vibration. It brought together experts from many different backgrounds to present and discuss their most recent work on hand-arm vibration. This conference aimed to develop a better understanding of the health risks resulting from vibration exposure in order to improve risk control measures. This event was intended for scientists, occupational physicians, epidemiologists, machine manufacturers, metrologists, health and safety practitioners, standardization groups, and government agencies.

Christophe Noël and Jacques Chatillon

Editors

Editorial

Statement of Peer Review [†]

Christophe Noël *[iD] and Jacques Chatillon

Institut national de recherche et de sécurité (INRS), 54519 Vandœuvre-lès-Nancy, France; jacques.chatillon@inrs.fr

* Correspondence: christophe.noel@inrs.fr; Tel.: +33-383-50-21-12

† All the papers published in the volume are presented at the 15th International Conference on Hand-Arm Vibration, Nancy, France, 6–9 June 2023.

In submitting conference proceedings to *Proceedings*, the volume editors of the proceedings certify to the publisher that all papers published in this volume have been subjected to peer review administered by the volume editors. Reviews were conducted by expert referees to the professional and scientific standards expected of a proceedings journal.

- Type of peer review: single-blind;
- Conference submission management system: dedicated website;
- Number of submissions sent for review: 46;
- Number of submissions accepted: 42;
- Acceptance rate (number of submissions accepted/number of submissions received): 91%;
- Average number of reviews per paper: 1.8;
- Total number of reviewers involved: 7;
- Any additional information on the review process: not applicable.

This publication collates the proceedings of the 15th International Conference on Hand-Arm Vibration, held from 6 to 9 June 2023 in Nancy, France. The conference was organized by the French Research and Safety Institute for the Prevention of Occupational Accidents and Diseases (INRS) under the patronage of the International Advisory Committee on Hand-Arm Vibration.

Organising Committee

INRS National Research and Safety Institute, France

- Dr. Jacques Chatillon
- Ms. Aline Marcelin
- Dr. Christophe Noël (chair)
- Mr. Bertrand Tinoco

Scientific Committee

- Dr. Emmanuelle Jacquet, FEMTO-ST University of Franche-Comté, France
- Dr. Kristine Krajnak, NIOSH National Institute for Occupational Safety and Health, USA
- Mr. Hans Lindell, RISE Research Institutes of Sweden, Sweden
- Dr. Christophe Noël, INRS National Research and Safety Institute, France
- Mr. Paul Pitts, Science Division, Health and Safety Executive, UK
- Prof. Nicla Settembre, CHRU Regional and University Hospital of Nancy, University of Lorraine, France
- Dr. Alice Turcot, INSPQ National Institute of Public Health of Quebec, Canada

International Advisory Committee on Hand-Arm Vibration

- Mr. Paul Pitts (Chair), Science Division, Health and Safety Executive, UK
- Dr. Christophe Noël (vice-chair) INRS National Research and Safety Institute, France
- Prof. Massimo Bovenzi, University of Trieste, Clinical Unit of Occupational Medicine, Italy
- Prof. Anthony Brammer, Envir-0-Health Solutions, Canada

Citation: Noël, C.; Chatillon, J. Statement of Peer Review. *Proceedings* **2023**, *86*, 42. https://doi.org/10.3390/proceedings2023086042

Published: 18 May 2023

- Dr. Renguang Dong, NIOSH National Institute for Occupational Safety and Health, USA
- Prof. Xing Gao, Capital Medical University, China
- Prof. Tatsuya Ishitake, Kurume University School of Medicine, Japan
- Mr. Uwe Kaulbars, Humanvibration – Mechanische Schwingungen, Germany
- Prof. Ronnie Lundstrom, Department of Public Health and Clinical Medicine, Umeå University, Sweden
- Prof. Setsuo Maeda, Engineering Department within the School of Science and Technology, Nottingham Trent University, UK
- Prof. Kazuhisa Miyashita, Department of Hygiene, School of Medicine, Wakayama Medical University, Japan
- Dr. Tohr Nilsson, Department of Public Health and Clinical Medicine, Umeå University, Sweden
- Dr. Ying Ye, Institute of Sound and Vibration Research, Southampton University, UK

Conflicts of Interest: The authors declare no conflict of interest.

Proceeding Paper

Physiological Effects of Single Shocks on the Hand-Arm System—A Randomized Experiment [†]

Elke Ochsmann [1,*], Alexandra Corominas [1], Uwe Kaulbars [2], Hans Lindell [3] and Benjamin Ernst [2]

[1] Department of Medicine, Institute of Occupational Medicine, University of Lübeck, 23562 Lübeck, Germany
[2] Department Ergonomics, Section Vibration, Institute for Occupational Safety and Health of the German Social Accident Insurance (IFA), 53757 St. Augustin, Germany
[3] Division Materials and Production, Research Institute Sweden (RISE), 43144 Mölndal, Sweden
* Correspondence: elke.ochsmann@uksh.de
[†] Presented at the 15th International Conference on Hand-Arm Vibration, Nancy, France, 6–9 June 2023.

Abstract: Physiological health effects (vibration perception thresholds and infrared skin temperature) of single-impact exposures and vibration exposures have been evaluated. In this experiment, a total of 52 healthy male participants were randomly exposed to single shocks of different frequencies ($1\ \text{s}^{-1}$, $4\ \text{s}^{-1}$, and $20\ \text{s}^{-1}$) and to random signal vibration exposures (4×5 min exposure duration). We observed frequency-dependent and eventually dose-dependent physiological effects. No exposure parameter systematically correlated to any of the examined physiological outcomes. This could hint at different pathways for physiological effects.

Keywords: single shock; randomization; physiological effect

1. Introduction

It is still unclear whether the same physiological or health effects can be expected for low frequency single-impact and vibration exposures [1–4]. This study investigated whether a change in the vibration perception threshold (VPT) and the surface skin temperature (T) can be detected after several single shock exposures of different frequencies ($1\ \text{s}^{-1}$, $4\ \text{s}^{-1}$, $20\ \text{s}^{-1}$) to the hand-arm system (4×5 min exposure duration). Furthermore, it was investigated whether the effects of single shock exposures can be compared with those of spectrum frequency exposures (random signal) of the same duration (and dose).

2. Materials and Methods

A total of 52 healthy male participants were randomly assigned to four experimental groups ($n = 13$ per group). Depending on the group, participants were exposed to either a 4×5 min single shock exposure of different frequencies ($1\ \text{s}^{-1}$, $4\ \text{s}^{-1}$, and $20\ \text{s}^{-1}$), followed by a random signal exposure, or a 4×5 min (+5 min) random signal exposure at the shaker. The participants stood upright during exposure and had their right hand positioned at the aluminum shaker handle with an angle of approx. 100° in the elbow joint. All other external test conditions (room temperature and humidity) were controlled and kept as constant as possible. Vibration perception thresholds (Vibrosense Meter II, VibroSense Dynamics, Malmö, Sweden) and infrared thermography (FLIR One Pro (FLIR Systems, Wilsonville, OR, USA) together with an iPhone 6 (Apple, Cupertino, CA, USA)) were used to detect early physiological effects. SPSS Version 28 (IBM, Armonk, NY, USA) was used for statistical analysis (descriptive analysis, Spearman correlation, and non-parametric tests). $p < 0.05$ was regarded as statistically significant.

Citation: Ochsmann, E.; Corominas, A.; Kaulbars, U.; Lindell, H.; Ernst, B. Physiological Effects of Single Shocks on the Hand-Arm System—A Randomized Experiment. *Proceedings* **2023**, *86*, 35. https://doi.org/10.3390/proceedings2023086035

Academic Editors: Christophe Noël and Jacques Chatillon

Published: 18 April 2023

3. Results

3.1. Vibration Perception Threshold (VPT)

In all frequency groups ($1\,s^{-1}$; $4\,s^{-1}$, $20\,s^{-1}$), there was a significant change in the VPT between baseline and post-exposure (1–5) measurement. However, the occurrence of the significant change is frequency- and dose-dependent. At a VPT test frequency of 125 Hz, (which seems to be more specific than higher test frequencies) a significant VPT increase occurs only after the 4th exposure sequence for $1\,s^{-1}$ shocks, after the 3rd exposure sequence for $4\,s^{-1}$ shocks, and after the 1st sequence for $20\,s^{-1}$ shocks. The results of the random signal exposure are similar to that of the $20\,s^{-1}$ exposure (Figure 1).

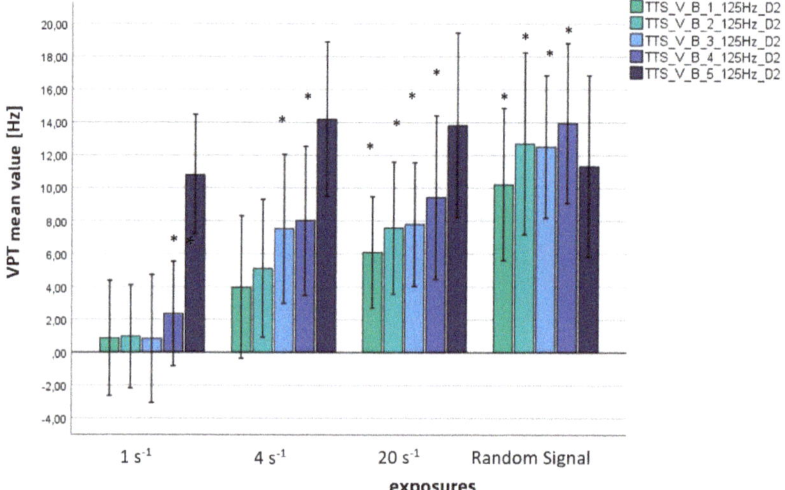

Figure 1. VPT (baseline exposures 1–5) at finger D2 of the exposed right hand (Hz: test frequency 125 Hz, bars: mean values, whiskers: 95% CI; N total = 52); *: $p < 0.05$.

3.2. Infrared Thermography (IR-T)

An overall decrease in the IR temperature of dorsal fingers of the exposed hand could be observed after four single shock exposures of the respective frequencies (exposures 1.0–4.1). In most fingers, this decrease was statistically significant. After four random signal exposures, no statistically significant differences in the IR dorsal finger temperature was observed. The development of the overall negative temperature gradient was based on several episodes of temperature loss during exposure and re-warming between exposures (see Figure 2).

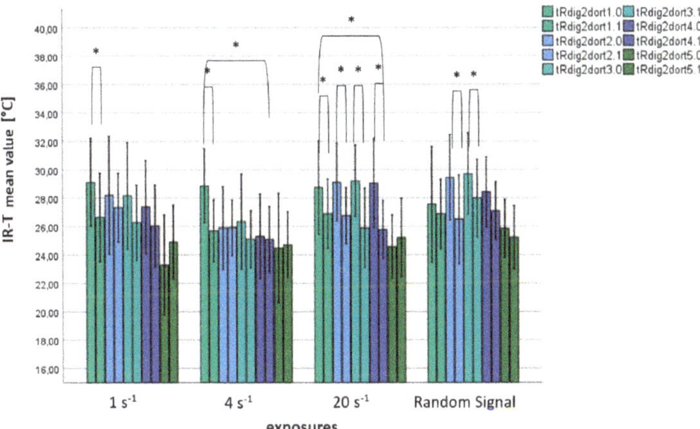

Figure 2. IR temperature measurements (pre and post exposure) at dorsal finger D2 of the exposed right hand (°C, bars: mean values, whiskers: 95% CI; N total = 52); exposure 1—exposure 5; *: $p < 0.05$.

3.3. Correlations between Exposure Parameters and VPT and IR-T

We could not identify repeating correlation patterns between exposure parameters at different frequencies and the examined physiological outcomes.

4. Discussion and Conclusions

Physiological effects after single shock exposure comprise temporary threshold shifts of vibration perception and skin temperature, similar to the expected responses caused by the hand-arm vibration in general. We found evidence that the (vascular and neurological) effects of single shock exposures might occur frequency- and dose-dependent, but also noticed that both endpoints do not follow the same patterns. We propose that this finding could be related to different causal pathways of the respective vascular and neurological endpoints. Different pathological mechanisms, to some extent, could also explain the lack of repeating patterns with regard to correlation between exposure and outcome parameters. All in all, our results suggest that early onset of physiological effects due to single shock exposures already occur below the existing exposure thresholds. While the prognostic value of these early physiological effects for the development and therefore prevention of the hand-arm vibration syndrome remain as yet unclear, further research is warranted to improve our understanding of underlying mechanisms responsible for single shock related health effects.

Author Contributions: Conceptualization, E.O., H.L. and U.K.; investigation, A.C. and B.E.; analysis, E.O. and B.E.; writing—original draft preparation, E.O.; writing—review and editing, all authors. All authors have read and agreed to the published version of the manuscript.

Funding: This research was funded by DGUV Forschungsförderung, FP-415 (www.dguv.de, accessed on 12 April 2023).

Institutional Review Board Statement: The study was conducted in accordance with the Declaration of Helsinki, and approved by the Ethics Committee of Luebeck University (protocol code 20-099, April 2020).

Informed Consent Statement: Informed consent was obtained from all subjects involved in the study.

Data Availability Statement: Not available.

Acknowledgments: We would like to thank all participants, and all technical staff for supporting this study.

Conflicts of Interest: The authors declare no conflict of interest.

References

1. Louda, R.; Rouskova, H.; Svoboda, L.; Muff, V. Diseases and disorders resulting from hand-arm shocks. In Proceedings of the 6th International Conference on Hand-Arm-Vibration, Bonn, Germany, 19–22 May 1992.
2. Schenk, H.; Heine, G. Effects of shocks on the peripheral nervous system. In Proceedings of the 6th International Conference on Hand-Arm-Vibration, Bonn, Germany, 19–22 May 1992.
3. Dandanell, R.; Engström, K. Vibration from riveting tools in the frequency range 6 Hz—10 MHz and Raynauds's phenomenon. *Scand. J. Work Environ. Health* **1986**, *12*, 338–342. [CrossRef] [PubMed]
4. Dupuis, H.; Schäfer, N. Effects of impulse vibration on the hand-arm system. *Scand. J. Work Environ. Health* **1986**, *12*, 320–322.

Proceeding Paper

Acute Vibrotactile Threshold Shifts in Relation to Force and Hand-Arm Vibration [†]

Shuxiang Gao and Ying Ye *

Human Factors Research Unit, Institute of Sound and Vibration Research, University of Southampton, Southampton SO17 1BJ, UK; s.gao@soton.ac.uk
* Correspondence: y.ye@soton.ac.uk
† Presented at the 15th International Conference on Hand-Arm Vibration, Nancy, France, 6–9 June 2023.

Abstract: Background: to investigate the acute effects of vibration and force level on vibrotactile perception following short-term exposure to hand-arm vibration (HAV). Methods: 12 individuals attended the test, and each of them completed a set of grasping tasks between 10 N and 80 N while being exposed to four intensities of HAV at 125 Hz, ranging from 5.5 to 44.0 m/s^2 (unweighted), for three minutes. The vibrotactile perception threshold (VPT, at 125 Hz) on the fingertip was assessed before and after the exposure. Results: Vibration caused considerable reductions in VPT, and the higher the HAV amplitude, the more the VPT shifted. There were also noticeable VPT shifts brought on by the force increase, but the force increase from 40 N to 80 N could not make more of a difference at higher vibration levels. Conclusions: Vibrotactile perception was sensitive to the vibration level, and was affected by the applied hand force when the vibration intensity was modest. With high vibration levels, the further sensorineural response to the force is limited after the force reaches a certain level.

Keywords: hand-arm vibration; grip exertions; vibrotactile perception; temporary threshold shift

1. Introduction

Employees who work with hand-held vibrating tools may experience health problems as a result of prolonged vibration exposure [1,2]. Their sensory nerves, blood vessels, muscles and joints could be damaged, causing considerable pain and even disability [3,4] Thus, it is worth investigating the long-term as well as short-term physiological effects of exposure to hand-arm vibration (HAV) for the purpose of preventing related diseases. Among the existing research methods, the measurement of the vibrotactile perception threshold (VPT) has been proposed as a useful technique for screening and diagnosis as it corresponds to the HAV-induced nerve damage that occurs at a relatively early stage [5–7]. In addition, the acute response to vibration exposure could be detected by a temporary shift in the VPT, which was found to be sensitive to the vibrating intensity [8,9]. Apart from the vibration of the tool, the applied force and its impact is another important consideration in exposure assessment. A firm grip not only prevents slipping from the tool [10], but also determines how much vibration energy enters the hand-arm system [11,12]. However, limited attention has been paid to whether the neural response is more responsive to the level of vibration or to the active forces applied. This study aimed to investigate the dependence of temporary nerve function impairment (temporary shift in vibrotactile perception) on exposure to four hand-arm vibrations at 125 Hz and four hand force exertions. It was hypothesized that greater vibration and greater hand forces on the vibrating handle would increase the risk of decline in neurological perception, with comparatively high vibration levels being more crucial for regulation than hand force.

Citation: Gao, S.; Ye, Y. Acute Vibrotactile Threshold Shifts in Relation to Force and Hand-Arm Vibration. *Proceedings* **2023**, *86*, 8. https://doi.org/10.3390/proceedings2023086008

Academic Editors: Christophe Noël and Jacques Chatillon

Published: 10 April 2023

2. Materials and Methods

An electrodynamic shaker producing vibrations along the z-axis was positioned horizontally as shown in Figure 1. A sinusoidal 125-Hz vibration was applied to the right hand through a handle fixed to this shaker. Subjects were required to take a relaxed upright seating position, grasping the handle with a bent-arm posture, while having their left forearms supported at the heart level. Four levels of hand-arm vibration (HAV) stimuli were used as exposure conditions, V = 5.5, 11, 22 and 44 in the unit m/s^2 (unweighted), together with the four levels of grip force applied: F = 10, 20, 40 and 80 in the unit N. Forces measured by the handle's Kistler force sensor were displayed on the screen in front, helping subjects maintain them at the desired levels.

Figure 1. The experimental set-up.

Up to 12 healthy subjects with no prior vibration history participated in the study. Each subject went through all 16 exposure conditions, each containing a period of three minutes of holding the handle. Immediately after the exposure, the participants were instructed to release their hands from the handle and undergo the tests of the VPT, followed by an adequate recovery period (over 20 min). Different exposure situations were conducted in a randomized manner on four separate days for each subject.

The test of the VPT was performed by *HVLab* Vibrotactile Perception Meter for around one minute at 125 Hz on the right index's fingertip. The surrounding contact force was set to 2 N. The elbow of each subject was allowed to rest on a supporter during the test.

3. Results

The baseline VPT calculated for all subjects on the right index fingers was at 0.183 m/s^2 at 125 Hz. The amount of change (TTS) relative to the resting VPT values was studied thereafter, indicating a reduced vibration sensation after the exposure. More shift is believed to be associated with a stronger adverse effect.

Data analysis was conducted using nonparametric tests in SPSS. The Wilcoxon signed rank test was performed between conditions with different exposures.

Figure 2 shows the distribution of TTS in vibratory sensation tested at 125 Hz on the exposed right index finger. After exposure to a 3 min vibration, all individuals experienced a transient rise in the sensorineural threshold. The median TTS was 0.266–0.633 m/s^2 for a small amount of HAV of 5.5 m/s^2 r.m.s. (unweighted) and went up to 1.188–3.391 m/s^2 with the high vibration intensity of 44 m/s^2 r.m.s. (unweighted). Significant differences in TTS can be identified for all four vibration settings regardless of the force level (p = 0.001–0.002, Wilcoxon). In response to an increase in the vibration level by a factor of two, the variations in threshold nearly doubled each time.

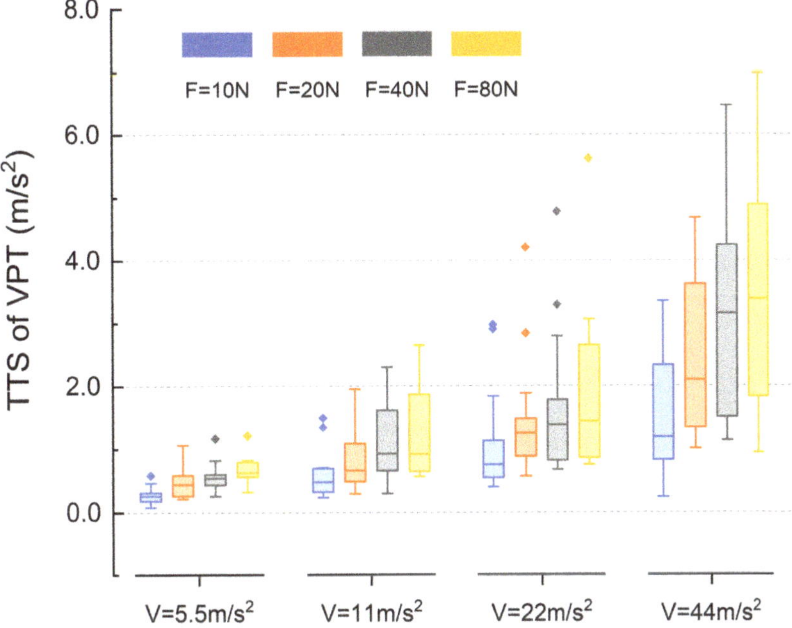

Figure 2. Distribution of the TTS in vibratory sensation as a function of force and vibration. Plotted median values (with interquartile range) of 12 subjects tested at 125 Hz on the right index finger after 3 min of exposure to different levels of grip force and vibration.

The impact on the TTS from the force level appears to follow a similar trend to that of the vibration level. In the presence of a vibration at 5.5 m/s^2 r.m.s. (unweighted), the reductions in perception were significantly different across the four force conditions (p = 0.001–0.006, Wilcoxon). The increase in hand forces was associated with the presence of more shifts in the VPT as shown in Figure 2, though the effect of twice as much force had less of an impact than that of twice as much vibration. At higher vibration levels ($\geq$11 m/s^2, unweighted), there are greater TTS differences with stronger hand forces (p = 0.001–0.006, Wilcoxon), except for the force increasing from 40 N to 80 N, which yielded similar results (p = 0.078–0.173, Wilcoxon).

4. Discussion

Generally, the finger perception thresholds of vibration tend to rise as a function of vibration intensity and hand force level. Exposure to vibration resulted in a clear loss in vibration sensation, which is consistent with the findings of earlier investigations [13]. The VPT results were sensitive to the vibration intensity regardless of the force level; the higher the HAV intensity is, the more vibrotactile perception at the fingertip shifts, which is the same finding as has been reported in previous studies [14,15]. The application of force during vibration exposure also affects the shift in the VPT. With the increase in force, the change in VPT follows an upward trend, similarly to the effect of the vibration level [8,12], suggesting that hand force may create a negative impact on hand sensorineural function. A similar influence of hand force on blood circulation during vibration was found in previous studies [16]. It is worth noting that a further increase in the large hand force was unable to introduce more of a reduction in vibration perception at high vibration levels, which indicates that the effect of force is limited, and may not be completely independent of that of vibration. This could be because the muscle tissue has already sustained the maximum deformation possible with a great vibration intensity, and because further increases in hand force will no longer be able to expose more receptors to the vibration level and cause additional sensation loss. In the case of both the vibration and force being at high levels, the VPT was still substantially affected by the vibration. The results strongly confirm the great influence of vibration on sensory nerves and underline that the regulatory role of the force applied should not be underestimated.

Author Contributions: Conceptualization, S.G. and Y.Y.; methodology, investigation, and original draft preparation, S.G.; supervision, review, and editing, Y.Y. All authors have read and agreed to the published version of the manuscript.

Funding: This research was funded by the Chinese Scholarship Council (CSC), grant no. 202008060219, under agreement [2020]71.

Institutional Review Board Statement: The study was approved by the Human Experimentation Safety and Ethics Committee at the University of Southampton (ERGO/FEPS/71800).

Informed Consent Statement: Informed consent was obtained from all subjects involved in the study.

Data Availability Statement: The data are not publicly available due to privacy reasons.

Acknowledgments: We thank Gary Parker and Peter Russell for their technical and instrumental support.

Conflicts of Interest: The authors declare no conflict of interest.

References

1. Griffin, M.J.; Erdreich, J. *Handbook of Human Vibration*; Acoustical Society of America: Melville, NY, USA, 1991.
2. Pelmear, P.; Wasserman, D. *A Comprehensive Guide for Occupational Health Professionals*; OEM Press: Beverly Farms, MA, USA, 1998.
3. Bovenzi, M. Medical aspects of the hand-arm vibration syndrome. *Int. J. Ind. Ergon.* **1990**, *6*, 61–73. [CrossRef]
4. Gemne, G. Diagnostics of hand-arm system disorders in workers who use vibrating tools. *Occup. Environ. Med.* **1997**, *54*, 90–95. [CrossRef] [PubMed]
5. Brammer, A.J.; Sutinen, P.; Diva, U.A.; Pyykkö, I.; Toppila, E.; Starck, J. Application of metrics constructed from vibrotactile thresholds to the assessment of tactile sensory changes in the hands. *J. Acoust. Soc. Am.* **2007**, *122*, 3732–3742. [CrossRef] [PubMed]
6. *ISO 13091-1: 2001*; Mechanical Vibration—Vibrotactile Perception Thresholds for the Assessment of Nerve Dysfunction—Part 1: Methods of Measurement at the Fingertips. International Organization for Standardization: Geneva, Switzerland, 2001.
7. Ye, Y.; Griffin, M.J. Assessment of thermotactile and vibrotactile thresholds for detecting sensorineural components of the hand-arm vibration syndrome (HAVS). *Int. Arch. Occup. Environ. Health* **2018**, *91*, 35–45. [CrossRef] [PubMed]
8. Löfgren, A.; Vihlborg, P.; Fornander, L.; Bryngelsson, I.-L.; Graff, P. Nerve Function Impairment After Acute Vibration Exposure. *J. Occup. Environ. Med.* **2020**, *62*, 124–129. [CrossRef] [PubMed]
9. Burström, L.; Lundström, R.; Hagberg, M.; Nilsson, T. Vibrotactile Perception and Effects of Short-Term Exposure to Hand-Arm Vibration. *Ann. Occup. Hyg.* **2009**, *53*, 539–547. [CrossRef] [PubMed]
10. Radwin, R.G.; Armstrong, T.J.; Chaffin, D.B. Power hand tool vibration effects on grip exertions. *Ergonomics* **1987**, *30*, 833–855. [CrossRef]
11. Gao, S.; Ye, Y. Acute Vascular Response of Hand to Force and Vibration. *Vibration* **2022**, *5*, 153–164. [CrossRef]
12. Hartung, E.; Dupuis, H.; Scheffer, M. Effects of grip and push forces on the acute response of the hand-arm system under vibrating conditions. *Int. Arch. Occup. Environ. Health* **1993**, *64*, 463–467. [CrossRef] [PubMed]
13. Malchaire, J.; Rodriguez Diaz, L.S.; Piette, A.; Gonçalves Amaral, F.; de Schaetzen, D. Neurological and functional effects of short-term exposure to hand-arm vibration. *Int. Arch. Occup. Environ. Health* **1998**, *71*, 270–276. [CrossRef] [PubMed]
14. Harada, N. Studies on the Changes in the Vibratory Sensation Threshold at the Fingertip in Relation to Some Physical Parameters of Exposed Vibration Part 2. A study on the equal TTS curves of the vibratory sensation and the hygienic allowable limit of portable mechanized tool. *Nippon Eiseigaku Zasshi Jpn. J. Hyg.* **1978**, *33*, 706–717.

15. Thonnard, J.-L.; Masset, D.; Penta, M.; Piette, A.; Malchaire, J. Short-term effect of hand-arm vibration exposure on tactile sensitivity and manual skill. *Scand. J. Work Environ. Health* **1997**, *23*, 193–198. [CrossRef] [PubMed]
16. Gao, S.; Ye, Y. Acute effects of force and vibration on finger blood flow. In Proceedings of the 8th American Conference on Human Vibration, online, 23–25 June 2021.

Proceeding Paper

Cold Response of Digital Vessels and Metrics of Daily Vibration Exposure †

Massimo Bovenzi [1,*] and Marco Tarabini [2]

1   Clinical Unit of Occupational Medicine, Department of Medical Sciences, University of Trieste, 34100 Trieste, Italy
2   Department of Mechanical Engineering, Politecnico di Milano, 20100 Milano, Italy; marco.tarabini@polimi.it
*   Correspondence: bovenzi@units.it; Tel.: +39-0403992340
†   Presented at the 15th International Conference on Hand-Arm Vibration, Nancy, France, 6–9 June 2023.

Abstract: The cold response of the digital arteries in a cohort of vibration-exposed workers was related to measures of daily vibration exposure expressed in terms of r.m.s. acceleration magnitude normalised to an 8-hour day, and frequency was weighted according to either the frequency weighting W_h defined in ISO 5349-1:2001 ($A_h(8)$ in ms^{-2} r.m.s) or the hand–arm vascular frequency weighting W_p proposed in the ISO Technical Report 18570:2017 ($A_p(8)$ in ms^{-2} r.m.s.). The metric $A_p(8)$, which assigns more weight to intermediate- and high-frequency vibrations (31.5–250 Hz), performed better for the prediction of cold-induced digital arterial hyperresponsiveness in the vibration-exposed workers than the measure $A_h(8)$ derived from the conventional ISO frequency weighting, which gives more importance to lower-frequency vibrations ($\leq$16 Hz).

Keywords: cold test; finger systolic blood pressure; frequency weighting; hand-transmitted vibration; vibration-induced white finger

Citation: Bovenzi, M.; Tarabini, M. Cold Response of Digital Vessels and Metrics of Daily Vibration Exposure. *Proceedings* **2023**, *86*, 6. https://doi.org/10.3390/proceedings2023086006

Academic Editors: Christophe Noël and Jacques Chatillon

Published: 10 April 2023

1. Introduction

Experimental studies have shown that the response of finger circulation to hand-transmitted vibration (HTV) is frequency-dependent [1]. Vibration frequencies $\geq$ 100 Hz can induce a stronger vasoconstriction than lower frequencies in either the human finger or animal models. Several epidemiological studies have reported that occupational exposure to intermediate- and high-frequency vibration is associated with an increased risk of a secondary form of Raynaud's phenomenon called vibration-induced white finger (VWF) [2]. These findings are in contrast with the frequency weighting W_h recommended by the ISO 5349-1 standard, which gives more weight to lower-frequency vibrations ($\leq$16 Hz) in the assessment of vibration-induced health disorders [3]. In the Italian arm of the EU VIBRISKS project [4], a supplementary hand–arm vascular weighting (W_p) proposed in the ISO/TR 18570 [5], which assigns more weight to intermediate- and high-frequency vibration (Figure 1), performed better than the ISO W_h curve for the prediction of the occurrence of subjective symptoms of VWF in a cohort of HTV workers.

The aim of the present study was to compare the relative performance of the vibration metrics constructed with either the frequency weighting W_h (ISO 5349-1) or the frequency weighting W_p (ISO/TR 18570) to predict, in addition to VWF symptoms, the cold response of the digital arteries in the VIBRISKS workers.

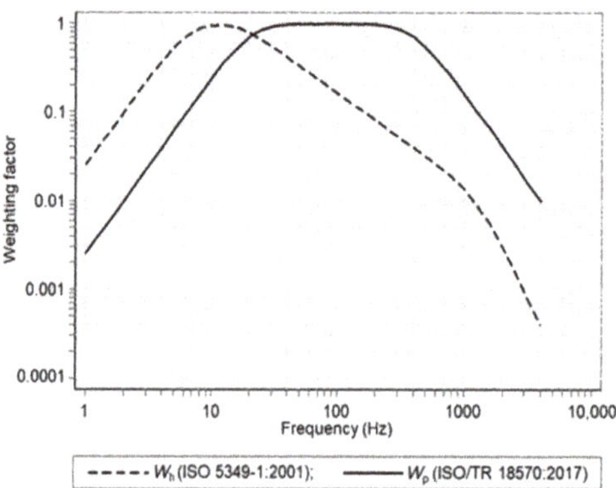

Figure 1. Comparison of frequency weighting functions (W) for hand-transmitted vibration. W_h: frequency weighting recommended in ISO 5349-1:2001; W_p: hand–arm vascular weighting defined in ISO/TR 18570:2017.

2. Material and Methods

The VIBRISKS cohort included 249 vibration-exposed workers (215 forestry operators and 34 stone workers) and 138 control men employed at the same companies and unexposed to HTV. They were investigated at baseline and annually in the autumn–winter seasons over a three-year follow-up period. The design of the VIBRISK prospective cohort study, the characteristics of the cohort workers, and the clinical criteria for the diagnosis of VWF symptoms have been described elsewhere [4].

2.1. The Cold Test

The cold test was carried out by means of a strain-gauge plethysmographic technique. The percentage change in finger systolic blood pressure (FSBP) from 30 to 10 °C (%FSBP$_{10°}$) in a test finger (FSBP$_t$), corrected for the change in systolic pressure in a reference finger (FSBP$_{ref}$) of the same hand, was calculated as follows:

$$\%\text{FSBP}_{10°} = (\text{FSBP}_{t,10°} \times 100)/[\text{FSBP}_{t,30°} - (\text{FSBP}_{ref,30°} - \text{FSBP}_{ref,10°})]\ (\%) \tag{1}$$

2.2. Vibration Exposure

Vibration measurements were obtained from the tools used by the forestry workers (chain saws, brush saws) and the stone workers (grinders, polishers, inline hammers) according to the recommendations of international standard ISO 5349-1 [3]. Triaxial (x, y, z) vibration magnitudes were measured as r.m.s. accelerations over the frequency range 1–4000 Hz using the frequency weightings W_h and W_p displayed in Figure 1.

The vibration total value (a_v) of the r.m.s. accelerations of tool i frequency weighted according to W_h or W_p (W_f) was calculated as follows:

$$a_{vi(W_f)} = \sqrt{a^2_{xi(W_f)} + a^2_{yi(W_f)} + a^2_{zi(W_f)}}\quad \left(\text{ms}^{-2}\ \text{r.m.s.}\right) \tag{2}$$

Daily vibration exposure was expressed in terms of r.m.s. acceleration magnitude normalised to an 8 h day ($A(8)$), and frequency-weighted according to W_h or W_p (W_f):

$$A(8)_{(W_f)} = \sqrt{\sum_{i=1}^{n} a^2_{vi(W_f)}\frac{T_i}{T_0}}\quad \left(\text{ms}^{-2}\ \text{r.m.s.}\right) \tag{3}$$

where a_v is the vibration total value of the r.m.s. acceleration of tool i, T_i is the duration of the ith operation with tool i in hours, and T_0 is the reference period of 8 h.

2.3. Data Analysis

Continuous variables were summarised with the median as a measure of central tendency and quartiles as a measure of dispersion. Comparison between unpaired data was carried out by means of non-parametric statistics. The relations of cold test outcome ($\%FSBP_{10°}$) to measures of daily vibration exposure expressed in terms of either $A_h(8)$ or $A_p(8)$ were estimated by maximum-likelihood random-effects regression models for repeated measures over the follow-up period. The Bayesian Information Criterion (BIC) was used to compare the fit of the regression models, including alternative measures of daily vibration exposure [6]. According to the strength of evidence rules for the difference (Δ) in BIC between models, ΔBIC 0–2 suggests no difference in the fit between models; ΔBIC 2–6 tends to give positive support for the model with the smaller BIC; ΔBIC 6–10 provides strong evidence for the model with the smaller BIC.

3. Results

The occurrence of symptoms of white finger over the study period was 7.2% in the controls and 21.7% in the HTV workers (17.7% in the forestry workers; 47.1% in the stone workers). There were no significant differences in age and anthropometric characteristics between groups, while current smoking was more prevalent among the HTV workers affected with VWF (Table 1). Daily vibration exposure in terms of either $A_h(8)$ or $A_p(8)$ was significantly greater in the VWF workers than in the HTV workers with no vascular symptoms ($p < 0.001$). Baseline FSBPs at 30 °C were similar in the controls and in the HTV workers with or without VWF, while the vasoconstrictor response to cold ($\%FSBP_{10°}$) was stronger in the VWF workers than in the controls and the non-VWF workers ($p < 0.0001$).

Table 1. Characteristics of the controls and the HTV workers. The results of the cold test are also shown. Data are given as medians (quartiles) or numbers (%). The HTV workers are divided into two sub-groups according to the occurrence of VWF over the study period.

Factors	Controls (n = 138)	HTV Workers (n = 249)	
		Non-VWF (n = 195)	VWF (n = 54)
Age (yr)	38.8 (34.1–45.9)	42.1 (33.6–46.8)	43.0 (34.8–52.2)
BMI (kg/m^2)	24.5 (23.0–27.2)	25.7 (23.2–27.4)	24.5 (23.2–26.8)
Current smokers (n)	29 (21.0)	85 (43.6)	28 (51.8) *
Drinkers (n)	104 (75.4)	145 (74.4)	47 (87.0)
$A_h(8)$ (ms^{-2} r.m.s.)	-	3.59 (2.48–5.21)	4.54 (3.44–7.94) **
$A_p(8)$ (ms^{-2} r.m.s.)	-	17.9 (12.5–27.4)	26.5 (16.1–78.9) **
Duration of exposure (y)	-	15 (7–21)	17 (11–23)
$FSBP_{t,30°}$ (mmHg)	120 (110–135)	130 (115–140)	125 (110–140)
$FSBP_{ref,30°}$ (mmHg)	130 (118–140)	130 (120–140)	130 (115–140)
$\%FSBP_{10°}$ (%)	92.9 (85.7–100)	91.7 (81.8–100)	81.7 (60.0–94.7) ***

χ^2 test: * $p < 0.001$; Mann–Whitney test (VWF vs. non-VWF workers): ** $p < 0.001$; Kruskal–Wallis test between groups: *** $p < 0.0001$.

The relation of cold test outcome ($\%FSBP_{10°}$) to measures of daily vibration exposure was investigated by means of two models with different sets of explanatory variables (Table 2). After excluding the controls from data analysis, in all models one unit of increase in $A_h(8)$ (1 ms^{-2}) or $A_p(8)$ (10 ms^{-2}) was significantly associated with an increase in the vasoconstrictor response of the digital vessels to cold (i.e., decrease in $\%FSBP_{10°}$). As expected, VWF symptoms were significantly related to the cold response of finger circulation. The BIC statistic suggests a better fit when $A_p(8)$ rather than $A_h(8)$ was included in the models as a predictor of vibration-induced digital vasoconstriction (ΔBIC 7 for both models).

Table 2. Relation of %FSBP$_{10°}$ to measures of daily vibration exposure expressed in terms of either $A_h(8)$ (ISO 5349-1) or $A_p(8)$ (ISO/TR 18570). Regression coefficients (95% CI) are estimated by maximum-likelihood random-effects regression models for repeated measures over the follow-up period. The likelihood ratio (LR) tests for the significance of the measures of daily vibration exposure and the Bayesian Information Criterion (BIC) for the comparison between models are shown.

Factors	Model 1 [a]		Model 2 [b]	
	$A_h(8)$ ($\times 1$ ms^{-2})	$A_p(8)$ ($\times 10$ ms^{-2})	$A_h(8)$ ($\times 1$ ms^{-2})	$A_p(8)$ ($\times 10$ ms^{-2})
$A_f(8)$ (ms^{-2} r.m.s.)	−1.23 (−1.63; −0.84)	−1.30 (−1.68; −0.92)	−0.98 (−1.38; −0.58)	−1.12 (−1.52; −0.71)
Duration of exposure (y)	-	-	−0.07 (−0.25; 0.12)	−0.03 (−0.22; 0.15)
VWF	-	-	−7.59 (−11.1; −4.03)	−7.02 (−10.6; −3.44)
LR test c^2 ($A_f(8)$) [c]	34.7	41.1	20.4	26.8
Model fitting (BIC)	7746	7739	7780	7773
ΔBIC	7		7	

[a] Adjusted by survey time and %FSBP$_{10°}$ at baseline. [b] Adjusted by age at entry, smoking, drinking, BMI, hand trauma or surgery, systemic disorders, daily use of medicines, leisure activity with vibrating tools, survey time, and %FSBP$_{10°}$ at baseline. [c] $p < 0.0001$ for $A_f(8)$ in both models.

4. Discussion

In this study, the metric $A_p(8)$ performed better than $A_h(8)$ for the assessment of the vasoconstrictor effect of cold in the digital arteries of HTV workers. This is consistent with our previous epidemiological findings that $A_p(8)$ was a better predictor of the occurrence over time of VWF symptoms in the VIBRISKS cohort compared to the measure of daily vibration exposure $A_h(8)$ recommended by ISO 5349-1 [4]. Vascular investigations have revealed that acute exposure to vibrations with equal frequency-weighted acceleration magnitudes can provoke a stronger reduction in the blood flow of the human finger for frequencies between 31.5 and 250 Hz compared with vibration at 16 Hz [7]. In animal models, exposure to high-frequency vibrations (250 Hz) was found to induce both functional (increased oxidative stress) and structural (arterial remodeling and narrowing) changes in the ventral tail arteries of rats [1]. The results of these pathophysiological and morphological investigations provide biological plausibility to the epidemiological findings of an increased occurrence of VWF symptoms in HTV workers operating power tools generating high-frequency vibration. Overall, the present study and our previous epidemiological surveys suggest that the evaluation of vibration exposure by means of a frequency weighting which assigns more weight to intermediate- and high-frequency vibration (31.5–250 Hz) is more appropriate for the assessment and the prediction of subjective symptoms and objective signs of vibration-related vascular disorders compared to the assessment method recommended by the current ISO 5349-1 standard, which tends to overestimate the vascular effects of lower-frequency vibration ($\leq$16 Hz).

Author Contributions: Conceptualization, M.B. and M.T.; methodology, M.B. and M.T.; software, M.B.; validation, M.B. and M.T.; formal analysis, M.B.; investigation, M.B.; resources, M.B.; data curation, M.B.; writing–original draft preparation, M.B.; writing–review and editing, M.B. and M.T.; visualization, M.B. and M.T.; supervision M.T.; project administration, M.B.; funding acquisition, M.B. All authors have read and agreed to the published version of the manuscript.

Funding: This research was supported by the European Commission under the Quality of Life and Management of Living Resources programme—Key Action 4–Environment and Health–Project No. QLK4-2002-02650 (VIBRISKS).

Institutional Review Board Statement: The study was conducted in accordance with the Declaration of Helsinki, and obtained ethical approval by the Local Health Authorities of the Italian NHS.

Informed Consent Statement: Informed consent was obtained from all subjects involved in the study.

Data Availability Statement: The data presented in this study are unavailable due to privacy.

Conflicts of Interest: The authors declare no conflict of interest.

References

1. Krajnak, K.; Riley, D.A.; Wu, J.; McDowell, T.; Welcome, D.E.; Xu, X.S.; Dong, R.G. Frequency-dependent effects of vibration on physiological systems: Experiments with animals and other human surrogates. *Ind. Health* **2012**, *50*, 343–353. [CrossRef] [PubMed]
2. Bovenzi, M. Epidemiological evidence for new frequency weightings for hand-transmitted vibration. *Ind. Health* **2012**, *50*, 377–387. [CrossRef] [PubMed]
3. *SO 5349-1*; Mechanical Vibration—Measurement and Evaluation of Human Exposure to Hand Transmitted Vibration—Part 1: General Requirements. ISO: Geneva, Switzerland, 2001.
4. Bovenzi, M.; Pinto, I.; Picciolo, F. Risk assessment of vascular disorders by a supplementary hand-arm vascular weighting of hand-transmitted vibration. *Int. Arch. Occup. Environ. Health* **2019**, *92*, 129–139. [CrossRef] [PubMed]
5. *ISO/TR 18570*; Mechanical Vibration—Measurement and Evaluation of Human Exposure to Hand Transmitted Vibration—Supplementary Method for Assessing Risk of Vascular Disorders. ISO: Geneva, Switzerland, 2017.
6. Raftery, A.E. Bayesian model selection in social research. *Sociol. Methodol.* **1995**, *25*, 111–163. [CrossRef]
7. Bovenzi, M.; Lindsell, C.J.; Griffin, M.J. Acute vascular responses to the frequency of vibration transmitted to the hand. *Occup. Environ. Med.* **2000**, *57*, 422–430. [CrossRef] [PubMed]

Proceeding Paper

Effects of Applied Pressure on Sensorineural and Peripheral Vascular Function in an Animal Model of Hand-Arm Vibration Syndrome †

Kristine Krajnak *, Christopher Warren, Xueyan S. Xu, Stacey Waugh, Phillip Chapman, Daniel E. Welcome and Ren G. Dong

Health Effects Laboratory Division, National Institute for Occupational Safety and Health, Morgantown, WV 26508, USA; cpw4@cdc.gov (C.W.); fze2@cdc.gov (X.S.X.); ztz6@cdc.gov (S.W.); ttf4@cdc.gov (P.C.); zzw8@cdc.gov (D.E.W.); rdk6@cdc.gov (R.G.D.)
* Correspondence: ksk1@cdc.gov; Tel.: +1-304-285-5964
† Presented at the 15th International Conference on Hand-Arm Vibration, Nancy, France, 6–9 June 2023.

Abstract: Hand-arm vibration syndrome (HAVS) is characterized by cold-induced vasospasms of peripheral vasculature, by changes in sensorineural function and by pain. Vibration frequency and amplitude along with pressure applied at the fingertips while gripping a tool may also affect vascular and sensorineural function. Little is known about how these two exposure factors interact to affect the risk of developing HAVS. This study uses a newly developed rat tail model to examine the effects of vibration on vascular and sensorineural function. Exposure to 2N of pressure for 10 consecutive days resulted in an increase in blood flow in the tail, which may have been the result of an increased sensitivity of the arteries to acetylcholine. There was also an increased sensitivity of the small myelinated fibbers to electrical stimulation and of the sensory receptors to a pressure stimulus. Based on these findings, pressure has its own effects on vascular and sensorineural physiology, and these effects can be different from those of vibration.

Keywords: hand-arm vibration; grip force; pressure; vascular function; sensorineural function

Citation: Krajnak, K.; Warren, C.; Xu, X.S.; Waugh, S.; Chapman, P.; Welcome, D.E.; Dong, R.G. Effects of Applied Pressure on Sensorineural and Peripheral Vascular Function in an Animal Model of Hand-Arm Vibration Syndrome. *Proceedings* 2023, 86, 15. https://doi.org/10.3390/proceedings2023086015

Academic Editors: Christophe Noël and Jacques Chatillon

Published: 11 April 2023

1. Introduction

Hand-arm vibration syndrome (HAVS) is characterized by changes in sensorineural and peripheral vascular function. The primary symptom of HAVS is cold-induced finger blanching due to spasms of the peripheral vasculature, which is referred to as vibration white finger (VWF). Other symptoms include a loss of sensory function and pain. Epidemiological, field and laboratory studies have examined the effects of vibration frequency and amplitude on the risk of developing the symptoms of HAVS [1–4]. However, grip force, or pressure applied at the finger tips while gripping a tool, may also affect vascular and sensorineural function [5,6].

Researchers at NIOSH have developed a new rat tail model for examining the effects of vibration and applied pressure on vascular and sensorineural function. A diagram of the new model can be seen in Figure 1, and a detailed description of the model is presented in [6]. This model was used to study the effects of applied pressure on physiological measures of vascular and sensorineural function.

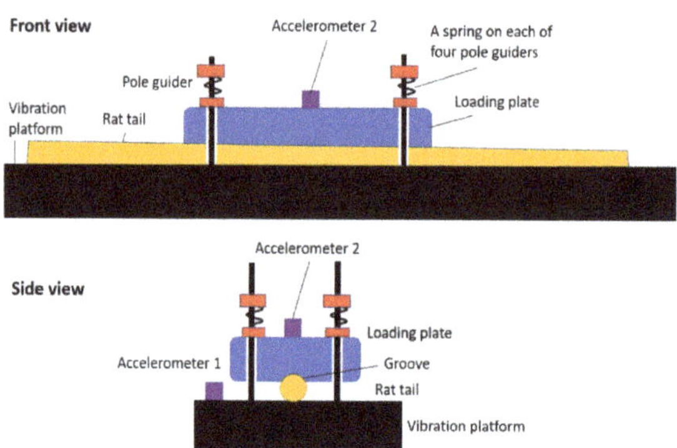

Figure 1. Modified rat tail model (Adapted from [7]).

2. Methods

2.1. Animals

Male Sprague Dawley rats (6 weeks of age; approximately 250 g) were obtained from Hilltop Breeders (Scottsdale, PA, USA) and acclimated to mild physical restraint in a Broome style restrainer. After acclimation, animals were assigned to two groups. One group was restrained and had their tails exposed to 2 newtons (N) of pressure ($n = 6$) for 4 h/day for 10 consecutive days. A second group (controls = 3) was maintained in restrainers without pressure applied to their tails. All procedures were approved by the Institutional Animal Care and Use Committee and were in compliance with both the Public Health Service Policy on Humane Care and Use of Laboratory Animals and the NIH Guide for the Care and Use of Laboratory Animals.

2.2. Physiological Measures

2.2.1. Laser Doppler

On days 1, 5 and 10 of the study, blood flow was measured by laser Doppler (Perimed: Järfälla, Sweden) both before and after exposure. Blood flow measurements were collected in perfusion units at 0.2 Hz for 5 min and average blood flow was calculated over for the 5 min recording period.

2.2.2. Sensory Measures

On days 2 and 9 of this study, sensory nerve function was measured using the current perception threshold test (CPT); moreover, on days 1, 5 and 10, sensitivity of receptors that respond to pressure was measured using the Randall–Selitto (R–S) test.

2.3. Microvessels

Animals were humanely euthanized 1 d following the last exposure by using an AVMA approved procedure, and ventral tail arteries were assessed for responsiveness to the α1-adrenoreceptor agonist, to phenylephrine (PE) and to the vasodilating agent, acetylcholine (ACh). Responsiveness to changes in internal vascular pressure was also measured.

2.4. Analyses

Average blood flow and average responses to the CPT or Randall-Sellitto test were calculated and used for analyses. Data were analysed using repeated measures with ANOVA and post-hoc *t*-tests. Pre-post measures were analysed using *t*-tests. $p < 0.05$ was considered significant.

3. Results

Post-exposure laser Doppler measurements were increased as compared to pre-exposure measurements in the animals exposed to pressure on day 5 of this study (Figure 2).

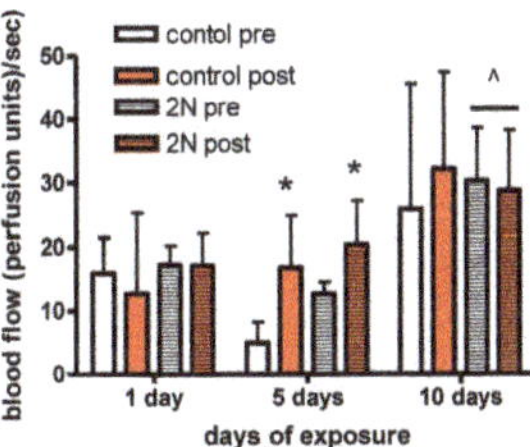

Figure 2. Exposure to control conditions or 2N of applied pressure resulted in acute increases in blood flow in both groups post exposure on day 5 of this experiment (* $p < 0.05$, different than pre-exposure measurements). There were no pre-post-exposure differences on day 10 of the experiment; however, in the pressure-exposed group, both pre- and post-exposure blood flow was increased in comparison with blood flow on days 1 and 5 (^ different than day 1, $p < 0.05$).

Pre-exposure laser Doppler measurements in the animals exposed to pressure were higher on day 10 of exposure (Figure 3).

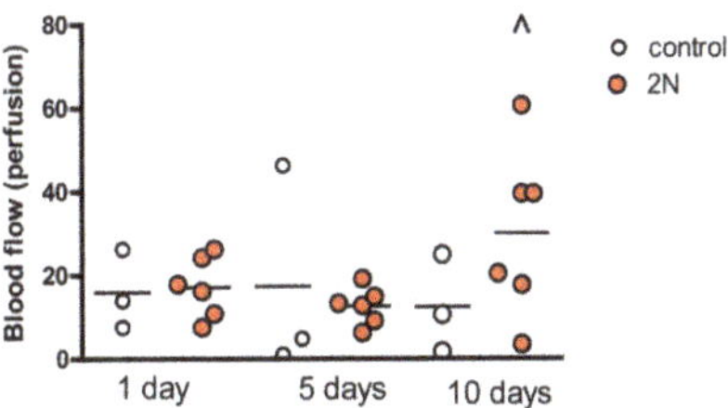

Figure 3. Pre-exposure blood flow was significantly increased after 10 days of exposure to 2N of pressure in comparison with blood flow in animals after 1 and 5 days of exposure (^ different than day 1, $p < 0.05$).

After 10 days of exposure, the responsiveness of the ventral tail artery to the vasoconstricting agent, PE (Figure 4A), and the vasodilation in response to the increasing vascular pressure (Figure 4C) were not affected in the arteries of the animals exposed to applied pressure. However, the arteries of the exposed animals were more sensitive to vasodilation in response to ACh (Figure 4B).

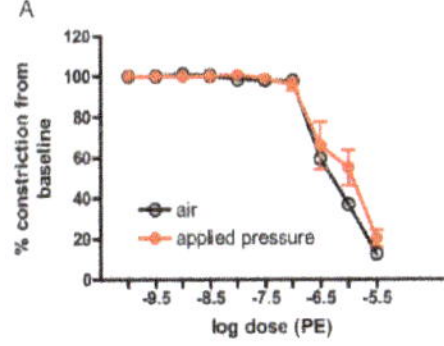
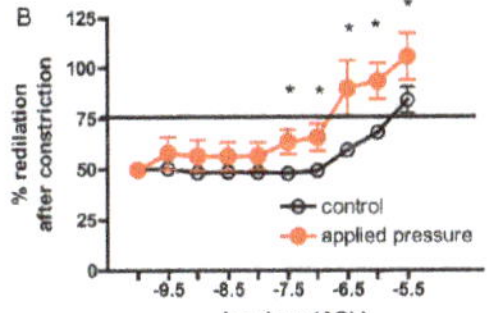
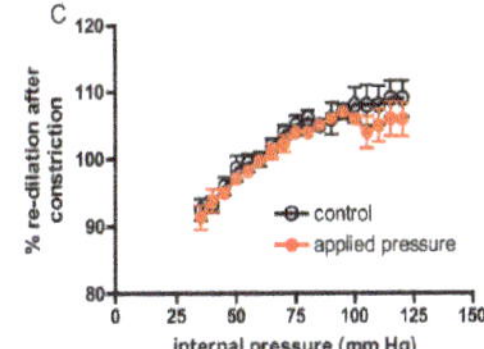

Figure 4. Sensitivity to PE-induced vasoconstriction in the ventral tail artery was not affected by exposure to pressure (A), but the artery was more sensitive to acetylcholine (ACh)-induced vasodilation (B). There were no exposure-related differences in vasodilation induced by increasing the internal pressure within the artery (C). (* different than controls, $p < 0.05$).

The 250 Hz pre-exposure CPT was reduced after 9 days of exposure to applied pressure (Figure 5).

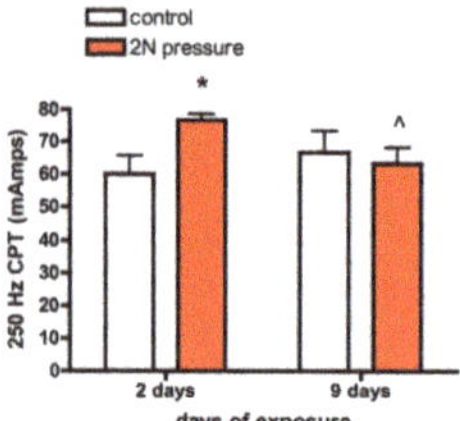

Figure 5. The (CPT) at 2000 and 5 Hz was not affected by exposure to pressure; however, after 9 days of exposure to pressure, animals were more sensitive to electrical stimulation at 250 Hz (reduction in threshold). (* different than same day control; ˆ less than 2 days of exposure, $p < 0.05$).

The tails of the animals exposed to pressure for 10 d were more sensitive to the pressure applied with an R–S pressure meter (Figure 6).

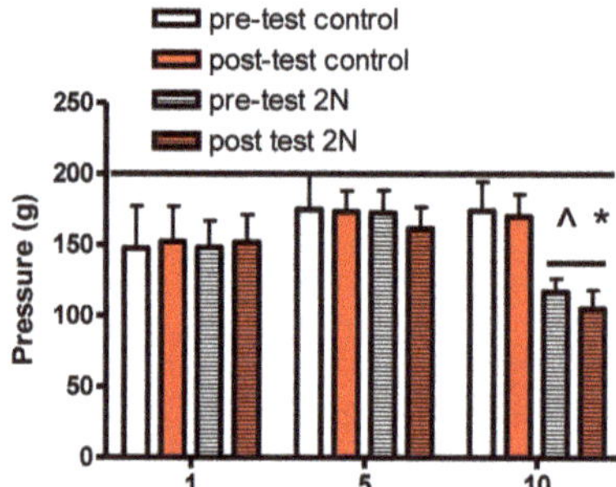

Figure 6. The R–S pressure meter was used to measure sensitivity to pressure applied in the exposed region of the tail (similar to the von Frey filament test). The line at 200 indicates the maximum amount of pressure that was applied (to prevent injury). There were no changes in the control group over time. Animals exposed to repeated pressure became more sensitive to applied stimulus after 10 days of exposure (ˆ different than day 1 same exposure; * different than same day pre-test, $p < 0.05$).

4. Discussion and Conclusions

- Blood flow was increased in the animals exposed to 2N of pressure (within the range of pressure generable by the fingertips of a tool user) for 10 days, suggesting that the blood vessels were *dilated.*
- The results of the vascular responsiveness studies were consistent with the results of the laser Doppler studies: pressure resulted in an increased responsiveness to ACh-induced re-dilation.
- The failure to find an increased response to internal pressure in the micro vessels studies (mimicking an increase in blood flow) in the exposed animals suggests that ACh or nitric oxide signalling may have been affected by the exposure.
- The CPT at 250 Hz was reduced in exposed animals, indicating an increase in sensitivity. These data suggest that pressure applied at the fingertips while gripping a tool may affect Aδ (small myelinated) the pressure and temperature sensing fibbers in the fingers.
- The findings from the Randall–Selitto test are consistent with the findings of the CPT, showing an increased sensitivity to sensory stimuli after repeated exposures.
- Additional studies are warranted to examine the effects of increased pressure and to determine the mechanisms underlying the changes in vascular responsiveness.

Author Contributions: Conceptualization, K.K., C.W., X.S.X., D.E.W. and R.G.D.; methodology, K.K., C.W., X.S.X., S.W., P.C. and R.G.D.; software, C.W. and X.S.X.; formal analysis, K.K.; investigation, K.K., P.C. and S.W.; resources and funding acquisition, this work was funded by a NIOSH NORA to K.K. and R.G.D.; data curation, K.K., P.C. and S.W.; writing original draft, K.K.; writing, review and editing, K.K., C.W., X.S.X., S.W., P.C., D.E.W. and R.G.D.; supervision, K.K. and R.G.D.; project administration, K.K. All authors have read and agreed to the published version of the manuscript.

Funding: This research was funded through the National Research Organization Agenda, NIOSH, grant 22037A9.

Institutional Review Board Statement: Not applicable.

Informed Consent Statement: Not applicable.

Data Availability Statement: Data will be available in NIOSHs study tracking and reporting system (STARS) once published.

Conflicts of Interest: The authors declare no conflict of interest.

References

1. Bovenzi, M. Epidemiological evidence for new frequency weightings of hand-transmitted vibration. *Ind. Health* **2012**, *50*, 377–387. [CrossRef] [PubMed]
2. Bovenzi, M.; Pinto, I.; Picciolo, F.; Mauro, M.; Ronchese, F. Frequency weightings of hand-transmitted vibration for predicting vibration-induced white finger. *Scand. J. Work. Environ. Health* **2011**, *37*, 244–252. [CrossRef] [PubMed]
3. Dong, R.G.; Welcome, D.E.; McDowell, T.W.; Wu, J.Z.; Schopper, A.W. Frequency Weighting Derived from Power Absorption of Fingers-Hand-Arm System under zh-Axis Vibration. *J. Biomec.* **2006**, *39*, 2311–2324. [CrossRef] [PubMed]
4. Krajnak, K.; Miller, G.R.; Waugh, S.; Johnson, C.; Li, S.; Kashon, M.; Wilder, D.; Rahmatalla, S.; Fathke, N. Vascular responses to vibration are frequency dependent. *J. Occup. Environ. Med.* **2010**, *52*, 584–594. [CrossRef] [PubMed]
5. Wu, J.Z.; Welcome, D.E.; McDowell, T.W.; Xu, X.S.; Dong, R.G. Modeling of the interaction between grip force and vibration transmissibility of a finger. *Med. Eng. Phys.* **2017**, *45*, 61–70. [CrossRef] [PubMed]
6. Starck, J.; Farkkila, M.; Aatola, S.; Pyykko, I.; Korhonen, O. Vibration Syndrome and Vibration in Pedestal Grinding. *Br. J. Ind. Med.* **1983**, *40*, 426–433. [CrossRef] [PubMed]
7. Dong, R.G.; Warren, C.; Xu, X.S.; Wu, J.Z.; Welcome, D.E.; Waugh, S.; Krajnak, K. A novel rat tail model for studying human finger vibration health effects. In Proceedings of the International Conference on Hand-Arm Vibration, Nancy, France, 6–9 June 2023.

Proceeding Paper

Development of a Novel Rat-Tail Model for Studying Finger Vibration Health Effects [†]

Ren G. Dong *[iD], Christopher Warren [iD], John Z. Wu, Xueyan S. Xu [iD], Daniel E. Welcome, Stacey Waugh and Kristine Krajnak

Physical Effects Research Branch, National Institute for Occupational Safety and Health, Morgantown, WV 26505, USA; cpw4@cdc.gov (C.W.); ozw8@cdc.gov (J.Z.W.); fze2@cdc.gov (X.S.X.); zzw8@cdc.gov (D.E.W.); ztz6@cdc.gov (S.W.); ksk1@cdc.gov (K.K.)
* Correspondence: rkd6@cdc.gov; Tel.: +1-304-285-6332
† Presented at the 15th International Conference on Hand-Arm Vibration, Nancy, France, 6–9 June 2023.

Abstract: The objective of this study was to develop a new rat-tail vibration model for studying finger vibration health effects. Unlike the previous rat-tail models, the vibration strain and stress of the rat tail can be conveniently and reliably quantified and controlled in a biological experiment. This makes it possible to identify and understand the quantitative relationships between these biodynamic responses and vibration health effects.

Keywords: vibration-induced white finger; hand-arm vibration syndrome; rat-tail vibration model

Citation: Dong, R.G.; Warren, C.; Wu, J.Z.; Xu, X.S.; Welcome, D.E.; Waugh, S.; Krajnak, K. Development of a Novel Rat-Tail Model for Studying Finger Vibration Health Effects. *Proceedings* **2023**, *86*, 24. https://doi.org/10.3390/proceedings2023086024

Academic Editors: Christophe Noël and Jacques Chatillon

Published: 13 April 2023

1. Introduction

The relationships between vibration biodynamic responses (vibration stress, strain, and power absorption density) and vibration biological effects are not sufficiently understood [1], probably because it is difficult to identify their exact relationship using human subjects or existing animal models. Systematically studying this fundamental scientific gap may help resolve major remaining issues outlined in the current ISO 5349-1 [2]: "the vibration exposures required to cause these disorders are not known precisely, neither with respect to vibration magnitude and frequency spectrum nor with respect to daily and cumulative exposure duration." As the first step is to help conduct systematic studies, the objective of this study was to develop a new rat-tail exposure system and a related analytical method to investigate combined health effects of vibration and contact pressure and to identify their relationships with biodynamic responses.

2. Methods

The proposed new rat-tail exposure system is shown in Figure 1. It is a modification of the existing NIOSH rat-tail model with an addition of a loading device [3]. The loading device is composed of a loading plate, a set of loading springs, and the fixtures for installing the device on the vibration platform.

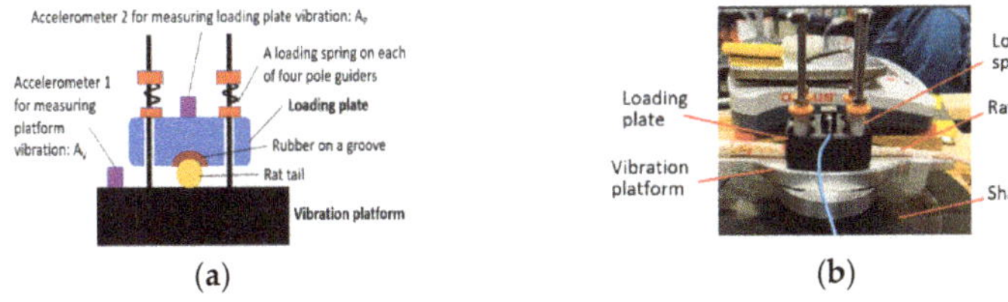

Figure 1. A novel rat-tail model: (**a**) model configuration; and (**b**) a pictorial view of the model.

The vibration platform can be installed on a shaker or a vibration tool. The loading plate is placed on the middle region (C12–18) of the rat tail that is constrained on the vibration platform. The static force is applied to the tail by loading the plate by compressing the loading springs. The loading plate has a groove with a conical taper, which can conform to the tapering diameter along the tail and secure the tail to the vibration platform.

The rat tail is likely to have a nonlinear force-deformation relationship. However, the dynamic properties of the tail can be locally linearized to reduce the complexity of the system analysis. Hence, the new rat-tail vibration-exposure system under several static forces (F_{PS}) was simulated using a linear analytical model illustrated in Figure 2.

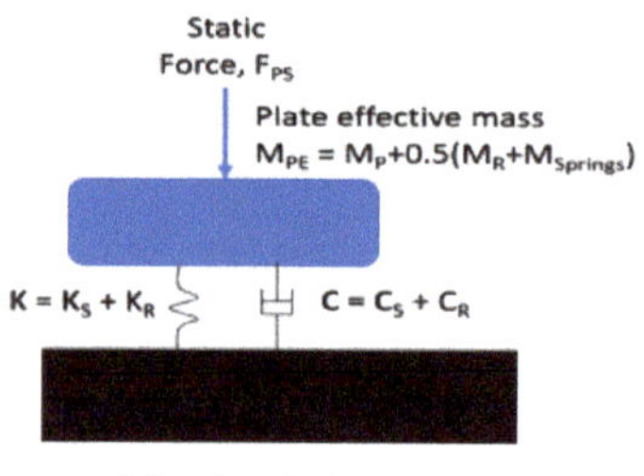

Figure 2. An analytical model of the new rat-tail vibration exposure.

The effective mass of the loading plate (M_{PE}) includes the loading plate mass (M_P), half of the mass of the loaded tail portion (M_R), and that of loading springs ($M_{Springs}$). The total stiffness of the exposure system (K) includes the tail stiffness (K_R) and loading spring stiffness (K_S). The total damping value includes the tail damping value (C_R) and that of the loading system (C_S).

Three biomechanical measures, including vibration stress, strain, and power-absorption density ($VPAD$), were considered for the quantification of the tail-vibration exposure. The average static or dynamic stress/pressure at the tail-plate interface was estimated from

$$\sigma_{Ave-static\ or\ dyn} \approx \frac{|F_{PS}\ \text{or}\ \mathbf{F}_{PD}|}{L_t \cdot b_t}, \tag{1}$$

where L_t is the plate length or tail contact length; b_t is the average tail contact width; $\mathbf{F}_{PD}$ is the complex dynamic force acting on the loaded portion of the tail. $\mathbf{F}_{PD}$ was estimated from

$$\mathbf{F}_{PD} \approx M_{PE}\mathbf{A}_P - [K_S(\mathbf{D}_P - \mathbf{D}_V) + C_S(\mathbf{V}_P - \mathbf{V}_V)], \tag{2}$$

where $\mathbf{D}_P$ and $\mathbf{D}_V$ are the complex vibration displacements of the loading plate and vibration platform, respectively; $\mathbf{V}_P$ and $\mathbf{V}_V$ are their vibration velocities; and $\mathbf{A}_p$ is the vibration acceleration of the loading plate.

The stress method assumes that the vibration stress at any point in the loaded portion of the tail (σ) is approximately proportional to the average contact stress. It also assumes that vibration injuries or biological effects are correlated with the stress magnitude and the number of stress cycles. Hence, the vibration exposure dose using the stress method (Γ_σ) for each frequency (f) was calculated from

$$\Gamma_\sigma \approx \sigma \cdot f^\alpha \cdot T = \Omega \cdot I_\sigma \cdot T, \tag{3}$$

where T is the time duration (seconds) of the rat-tail vibration exposure; α is termed as stress frequency weighting; Ω is the stress-proportional factor; and I_σ is termed as vibration-stress-dose index, which can be expressed as follows:

$$I_\sigma = \frac{\left| M_{PE}\mathbf{T}_P - \left[\frac{K_S}{(2\pi f)^2} + \frac{jC_S}{2\pi f}\right](1 - \mathbf{T}_P)\mathbf{A}_V \right|}{b_t \cdot L_t} f^\alpha, \tag{4}$$

where $j = \sqrt{-1}$; $\mathbf{T}_P$ is the transfer function of the loading plate ($=\mathbf{A}_P/\mathbf{A}_V$), where $\mathbf{A}_V$ is the vibration acceleration of the platform.

The strain is a measure of the material relative deformation. The maximum static strain of the tail subjected to a static force was estimated from the maximum compression deformation (Δ) and the average diameter of the loaded-tail portion (d_t):

$$\sigma_{\text{Maxi-static}} \approx \Delta/d_t, \tag{5}$$

The average dynamic strain for a given static force was expressed as follows:

$$\varepsilon_{Ave-dyn} \approx \frac{|\mathbf{D}_P - \mathbf{D}_V|}{h_t} = \frac{\left|\frac{\mathbf{A}_V - \mathbf{A}_P}{\omega^2}\right|}{h_t}, \tag{6}$$

where h_t is the deformed height of the loaded tail loaded with a static force, which can be measured in a static test.

Similar to the stress method, the strain method assumes that the distributed strain at any point (ε) is approximately proportional to the average dynamic strain, and the vibration injuries or biological effects are correlated with the strain and the number of strain cycles. Hence, the vibration exposure dose using the strain method (Γ_ε) can be written as follows:

$$\Gamma_\varepsilon \approx \varepsilon \cdot f^\beta \cdot T = Q \cdot I_\varepsilon \cdot T, \tag{7}$$

where β is termed as strain frequency weighting; Q is the strain-proportional factor; and I_ε is termed as the vibration-strain-dose index, which is expressed as follows:

$$I_\varepsilon = \frac{|\mathbf{A}_V - \mathbf{A}_P|}{\omega^2}\frac{f^\beta}{h_t} = \frac{|(1 - \mathbf{T}_P)\mathbf{A}_V|}{4\pi^2}\frac{f^{\beta-2}}{h_t}, \tag{8}$$

Similarly, with the model shown in Figure 2, the VPAD method for calculating the dose can be derived as follows:

$$\Gamma_{VPAD} \approx P \cdot \frac{\mathbf{C_R} \cdot |\mathbf{V}_P - \mathbf{V}_V|^2}{v} \cdot T = P \cdot I_{VPAD} \cdot T, \tag{9}$$

where v is the volume of the loaded tail portion ($\approx \pi \cdot (0.5d_t)^2 \cdot L_t$); P is the VPAD proportional factor; and I_{VPAD} is termed as VPAD dose index:

$$I_{VPAD} = \frac{C_R \cdot \left[\frac{|(\mathbf{T}_P - 1)\mathbf{A}_V|}{2\pi f}\right]^2}{v}, \tag{10}$$

A spring-calibration test, a damping test, and a rat-tail test were performed in this study to determine the system parameters, evaluate the feasibility of the new rat-tail model, and explore the characteristics of the exposure-dose indexes. The rat-tail test used a dissected tail obtained from a cadaver rat (the tissue collected post-mortem from the rat in a control group of another study).

3. Preliminary Results

As an example, the vibration transmissibility (the magnitude of transfer function) measured with 2.23 N static force is plotted in Figure 3a. The vibration-exposure-dose indexes for the three methods were calculated from the transfer function. While the stress and strain frequency weightings (α and β) cannot be determined in this study, they were assumed to be at unity (or 1.0) in the calculation to identify the fundamental characteristics of the indexes. To directly compare the index spectra, each index spectrum was normalized with respect to its maximum value. The results are plotted in Figure 3b. Their resonant frequencies were marginally higher than the transmissibility-resonant frequency. To explore the influences of the static force on the index spectra, the index spectrum was normalized with respect to the maximum value of the indexes for the same method under three forces (0.46, 1.28, and 2.23 N). The results for stress and strain methods are plotted in Figure 3c,d. The influences on the VPAD method were similar to those on the strain method.

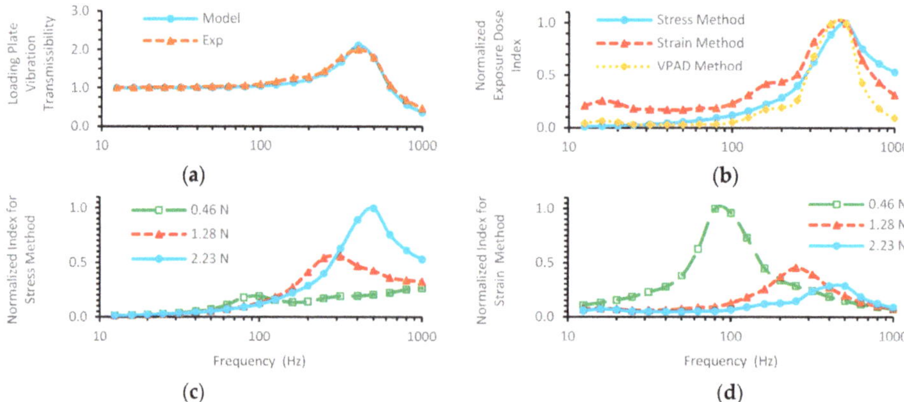

Figure 3. Preliminary results: (**a**) Measured and modeled vibration transmissibility spectra; (**b**) Normalized dose-index spectra of the three methods for 2.23 N static force; (**c**) Normalized stress-dose index for different static forces; (**d**) Normalized strain-dose index for different static forces.

4. Discussion and Conclusions

The vibration response of the new rat-tail model worked as expected, as demonstrated by the excellent agreement between the modeling and experimental results (Figure 3). As the tail mass of the loaded portion (about 0.21 g) is much less than that of the loading plate (>20 g), the dynamic force acting on this portion of the tail during a vibration exposure results primarily from the plate acceleration ($\approx M_P \mathbf{A}_P$). In this way, the biodynamic responses, such as vibration stress and strain of the tail tissues, can be effectively controlled and quantified by measuring and controlling the loading plate acceleration ($\mathbf{A}_P$). This makes it convenient and reliable to quantify the tail biodynamic responses required to investigate their relationships with biological effects.

The new methodology opens a window to view the mechanisms of the vibration health effects and their influencing factors. For example, the stress- and strain-dose index formulas (Equations (4) and (8)) suggest that the frequency may affect the vibration injuries or biological effects in two different ways: (I) the frequency determines the stress and strain magnitudes; they reach their peak values at the fundamental resonant frequency under each static force, as shown in Figure 3c,d; and (II) the frequency determines the number of cycles per second. These two different roles may result in different biological effects and/or their different frequency dependencies.

Another unique feature of the new rat-tail model is that the static force applied on the tail can be reliably controlled, which makes it possible to investigate the influences of the combined force and vibration exposures on the biological effects. The results presented in Figure 3 suggest that the static force affected the vibration stress- and strain-exposure doses differently. This is because the rat tail has a nonlinear mechanical behavior. At a low static force, the tissue stiffness is less than that at a higher static force; as a result, the same dynamic force results in a larger strain at the low static force than that at the higher static force. On the other hand, the vibration stress is independent of the deformation, but it depends primarily on the dynamic force itself (Equation (4)). This finding may help identify and understand the static-force influences on the vibration health effects and their frequency dependencies.

Overall, the new rat-tail model provides a new and effective tool for studying human finger vibration health effects.

Author Contributions: Conceptualization, R.G.D. and J.Z.W.; methodology, R.G.D. and C.W.; software, R.G.D. and X.S.X.; validation, J.Z.W., D.E.W. and K.K.; experiments and investigation, all authors; data curation, X.S.X.; writing—original draft preparation, R.G.D.; writing—review and editing, K.K., D.E.W. and S.W.; project administration, K.K. All authors have read and agreed to the published version of the manuscript.

Funding: This research received no external funding.

Institutional Review Board Statement: Not applicable.

Informed Consent Statement: Not applicable.

Data Availability Statement: The experimental data are available upon request.

Conflicts of Interest: The authors declare no conflict of interest.

References

1. Dong, R.G.; Wu, J.Z.; Xu, X.S.; Welcome, D.E.; Krajnak, K. A Review of Hand-Arm Vibration Studies Conducted by US NIOSH since 2000. *Vibration* **2021**, *4*, 482–528. [CrossRef]
2. *ISO 5349-1*; Mechanical Vibration—Measurement and Evaluation of Human Exposure to Hand-Transmitted Vibration—Part 1: General Requirements. International Organization for Standardization: Geneva, Switzerland, 2001.
3. Krajnak, K.; Waugh, S.; Miller, G.R.; Johnson, C.; Kashon, M.L. Vascular responses to vibration are frequency dependent. *J. Occup. Environ. Med.* **2010**, *52*, 584–594. [CrossRef] [PubMed]

proceedings

Proceeding Paper

Biomarkers in Patients with Hand-Arm Vibration Injury Entailing Raynaud's Phenomenon and Cold Sensitivity, Compared to Referents [†]

Eva Tekavec [1,*], Tohr Nilsson [2], Lars B. Dahlin [3], Anna Axmon [1], Catarina Nordander [1], Jakob Riddar [1] and Monica Kåredal [1]

[1] Division of Occupational and Environmental Medicine, Department of Laboratory Medicine, Lund University, 221 00 Lund, Sweden; anna.axmon@med.lu.se (A.A.); catarina.nordander@med.lu.se (C.N.); jakob.riddar@med.lu.se (J.R.); monica.karedal@med.lu.se (M.K.)

[2] Division of Sustainable Health and Medicine, Department of Public Health and Clinical Medicine, Umeå University, 901 87 Umeå, Sweden; tohr.nilsson@gmail.com

[3] Department of Translational Medicine—Hand Surgery, Lund University, 205 02 Malmoe, Sweden; lars.dahlin@med.lu.se

* Correspondence: eva.tekavec@med.lu.se

† Presented at the 15th International Conference on Hand-Arm Vibration, Nancy, France, 6–9 June 2023.

Abstract: The clinical evaluation of patients with neurosensory injury is based on quantitative sensory testing. Such tests require the patient's cooperation, which may sometimes hinder a correct diagnosis. Objective findings, e.g., with biomarkers, would therefore be valuable. We evaluated serum biomarkers of vascular and neural injury in 92 patients with vibration injuries and in 64 referents. Thrombomodulin (TM), a biomarker for endothelial damage, was elevated in HAVS patients with Raynaud's phenomenon (RP) compared to those without, as it also was in comparison to the referents. In the patients without RP, those with increased cold sensitivity showed somewhat higher—but not significantly different—values of TM ($p = 0.4$) than those without increased cold sensitivity, indicating an endothelial dysfunction or injury.

Keywords: hand-arm vibration syndrome; hand arm vibration injury; Raynaud's phenomenon; cold sensitivity; biomarkers; thrombomodulin; glial fibrillary acidic protein; endothelial dysfunction

Citation: Tekavec, E.; Nilsson, T.; Dahlin, L.B.; Axmon, A.; Nordander, C.; Riddar, J.; Kåredal, M. Biomarkers in Patients with Hand-Arm Vibration Injury Entailing Raynaud's Phenomenon and Cold Sensitivity, Compared to Referents. *Proceedings* **2023**, *86*, 27. https://doi.org/10.3390/proceedings2023086027

Academic Editors: Christophe Noël and Jacques Chatillon

Published: 14 April 2023

1. Introduction

The pathophysiology of hand-arm vibration syndrome (HAVS) is not fully clear. In most cases neurosensory symptoms proceed vascular ones [1]. Data suggest there is an individual susceptibility to HAVS, possibly due to contributions from coexisting medical conditions, for e.g., diabetes neuropathy [2]. Different pathophysiological mechanisms such as localized structural injuries affecting peripheral nerves and blood vessels and systemic inflammatory processes may exist [3–7]. Elevated plasma levels of thrombomodulin (TM), a marker for endothelial damage, are higher in vibration-exposed than in non-exposed workers [8,9]. Glial fibrillary acidic protein (GFAP), a proposed marker for axonal damage [10], has been detected in the nerve biopsies of type 2 diabetes subjects and controls [11], and elevated serum levels of GFAP correlate to decreased nerve action potentials [12]. In addition to symptoms of numbness, having tingling or not, having an impaired perception of touch, vibration, and temperature, and being diagnosed with Raynaud's phenomenon (RP), the patients often describe increased cold sensitivity [13]—a phenomenon that can mechanistically be induced by a vascular or neural pathway following vibration exposure [14]. Individuals with nerve injury may recover from increased cold sensitivity, whereas patients with HAVS may not [15].

Biomarkers for neural and vascular injuries may shed light on mechanisms as well as be used as a complemental objective method for diagnosis, since clinical assessment of the neurosensory component of HAVS mostly relies on quantitative sensory tests requiring cooperation of the evaluated patients.

Our aims were to investigate serum levels of biomarkers for vascular (TM) and for nerve (GFAP) injury in patients with neurosensory HAVS, with or without RP, compared to referents, as well as to study the association of increased cold sensitivity with the levels of TM and GFAP in HAVS patients without RP.

2. Materials and Methods

2.1. Study Group

Ninety-two patients diagnosed with neurosensory manifestations of HAVS and/or RP were recruited from our department of occupational and environmental medicine, together with 64 referents, who were without a vibration injury.

2.2. Questionnaires

Data on descriptive characteristics and symptoms were collected via a questionnaire (Table 1). Also, questions about job tasks and vibration exposure was assessed, not presented here.

Table 1. Descriptive characteristics of 92 patients with hand-arm vibration syndrome and 64 referents.

	Patients with Raynaud's Phenomenon (n = 45)	Patients without Raynaud's Phenomenon (n = 47)	Referents (n = 64)
Age [years; median (range)]	45 (24–64)	45 (21–64)	43 (26–62)
Females [n (%)]	1 (2)	5 (11)	9 (14)
Ongoing nicotine use [n (%)]	23 (51)	17 (36)	25 (40)
Previous frost bites in hands [n (%)]	6 (15) [a]	4 (9) [b]	3 (5) [c]
Raynaud's phenomenon [n (%)]	45 (100)	0 (0)	6 (10)
Cold sensitivity [n (%)]	44 (98)	36 (77)	10 (16)
Impaired perception of touch [n (%)]	29 (64)	16 (34)	10 (16)

[a] missing data from 4 patients; [b] missing data from 3 patients; [c] missing data from 4 referents.

2.3. Perception of Touch

Perception of touch was tested with a Semmes Weinstein Monofilament. If participants were unable to detect stimulation with a filament, No. 3.61, corresponding to 0.4 g of force, the perception of touch was considered impaired.

2.4. Biomarkers

Blood samples were collected in the morning. Serum was removed and stored at −80 °C until analysis. TM and GFAP were measured by using commercially available ELISA assays (TM by ELISA kit from BioVendor (Brno, Czech Republic); GFAP from Proteintech, (Manchester, UK)) according to instructions provided by the manufacturers. Samples were diluted in a ratio of 1:2 in sample diluent, and after sample preparation the optical density was recorded in a microplate reader at a wavelength of 450 nm. Duplicate readings were made. The detection limits (LOD) were 0.625 ng/mL for TM and 31 ng/mL for GFAP.

2.5. Statistics

The Kruskal–Wallis test was used to compare median values between the group of patients with and without RP and referents. Mann–Whitney U tests were used for post hoc analyses, and between the subgroups of patients with and without increased cold sensitivity. p-values below 0.05 were considered statistically significant. The results have not been adjusted for confounding effects.

3. Results

3.1. Biomarkers in Patients with and without RP and Referents

The Kruskal–Wallis test showed a statistically significant difference in TM levels between groups. Post hoc analyses showed a significant increase in the serum levels of TM in patients with RP compared to patients without RP, as well in patients with RP compared to the referents (Table 2). TM levels were highest in patients with RP and lowest in the referents.

Table 2. Biomarkers in patients with and without Raynaud's phenomenon and referents.

	Patients with Raynaud's Phenomenon (*n* = 45) Median (Range)	Patients without Raynaud's Phenomenon (*n* = 47) Median (Range)	Referents (*n* = 64) Median (Range)	*p*-Value
S-TM (ng/mL)	6.1 (2.7–30) [a, b]	5.2 (2.3–39) [a]	4.3 (<LOD–40) [b]	0.003
S-GFAP (pg/mL)	LOD [c] (LOD–3070)	LOD (LOD–2528)	LOD (LOD–3376)	0.28

[a] p = 0.02; [b] $p \leq 0.001$, [c] LOD = below lowest detection level. *p*-values from Kruskal–Wallis test. Post hoc analyses with Mann–Whitney U test.

3.2. Biomarkers in Patients without RP and with and without Increased Cold Sensitivity

Patients without RP, but reporting increased cold sensitivity, showed a somewhat higher but not significant increase in the serum levels of TM (median 5.2; range 2.3–39) compared to patients not reporting this symptom (4.6; 2.8–11); p = 0.4, Figure 1. For GFAP, there were no significant differences between any of the groups.

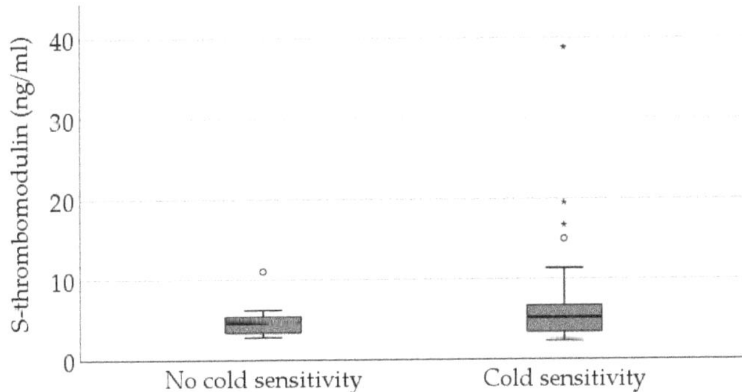

Figure 1. S-thrombomodulin in relation to increased cold sensitivity among HAVS patients without Raynaud's phenomenon. p = 0.4 (Mann–Whitney U test). Far outliers marked with a star and outliers marked with a circle.

4. Discussion

Patients with HAVS neurosensory injury and with RP had higher levels of TM than vibration-injured patients without RP. Patients with HAVS neurosensory injury and with RP also had higher levels of TM than the referents. There was no statistically significant difference between HAVS patients without RP and the referents. Increased levels of TM have been shown in previous studies on vibration-injured patients [8,9]. In one of the studies, however, there was no significant difference between patients with RP and those without RP [9]. This is in contrast to our findings that rather indicate that the RP symptoms is required to detect a significant change in the biomarker TM in HAVS patients. Patients with HAVS neurosensory injury without RP, reporting increased cold sensitivity, revealed higher levels of TM (although not significant) compared to those not reporting this symptom. Since TM is an endothelial marker, this may indicate that increased cold sensitivity originates from an endothelial dysfunction or injury which possibly could progress to RP.

The levels of GFAP did not differ significantly between the groups in this study, possibly because the neuropathy of the patients did not originate from any nerve injury involving GFAP or possibly because the samples were collected at a time point not reflecting the event for which the GFAP was released into the serum.

Author Contributions: Conceptualization, E.T. and C.N.; methodology, M.K.; formal analysis, E.T., C.N. and M.K.; data curation, E.T. and C.N. writing—original draft preparation, E.T., C.N. and M.K.; writing—review and editing, E.T., C.N., M.K., L.B.D., A.A., J.R. and T.N.; supervision, T.N. and L.B.D.; funding acquisition, C.N. All authors have read and agreed to the published version of the manuscript.

Funding: The study was funded by AFA Insurance (Grant number 170124).

Institutional Review Board Statement: The study was conducted in accordance with the Declaration of Helsinki and approved by the Regional Ethics Board in Lund. Sweden (No. 2018/15).

Informed Consent Statement: Participants gave informed consent to participate in the study before taking part.

Data Availability Statement: The datasets presented in this article are not readily available because public access to data is restricted to Swedish Authorities (Public Access to Information and Secrecy Act), but data can be available for researchers after a special review that includes approval of the research project by both an Ethics Committee and the Authorities' Data Safety Committees.

Acknowledgments: The authors would like to thank Elizabeth Huynh for skilful analyses of biomarkers and Ulla Andersson, Else Åkerberg Krook, Eva Assarsson, and Anna Larsson for the collection of the blood samples and handling of the data. We would also like to thank all participants in the patient, as well as the referent group.

Conflicts of Interest: The authors declare no conflict of interest.

References

1. Nilsson, T.; Wahlstrom, J.; Burstrom, L. Hand-arm vibration and the risk of vascular and neurological diseases-A systematic review and meta-analysis. *PLoS ONE* **2017**, *12*, e0180795. [CrossRef] [PubMed]
2. Dahlin, L.B.; Sanden, H.; Dahlin, E.; Zimmerman, M.; Thomsen, N.; Bjorkman, A. Low myelinated nerve-fiber density may lead to symptoms associated with nerve entrapment in vibration-induced neuropathy. *J. Occup. Med. Toxicol.* **2014**, *9*, 7. [CrossRef] [PubMed]
3. Stoyneva, Z.; Lyapina, M.; Tzvetkov, D.; Vodenicharov, E. Current pathophysiological views on vibration-induced Raynaud's phenomenon. *Cardiovasc. Res.* **2003**, *57*, 615. [CrossRef] [PubMed]
4. Kiedrowski, M.; Waugh, S.; Miller, R.; Johnson, C.; Krajnak, K. The effects of repetitive vibration on sensorineural function: Biomarkers of sensorineural injury in an animal model of metabolic syndrome. *Brain Res.* **2015**, *1627*, 216–224. [CrossRef] [PubMed]
5. Krajnak, K.; Waugh, S. Systemic Effects of Segmental Vibration in an Animal Model of Hand-Arm Vibration Syndrome. *J. Occup. Environ. Med.* **2018**, *60*, 886–895. [CrossRef] [PubMed]
6. Takeuchi, T.; Futatsuka, M.; Imanishi, H.; Yamada, S. Pathological changes observed in the finger biopsy of patients with vibration-induced white finger. *Scand. J. Work Environ. Health* **1986**, *12*, 280–283. [CrossRef] [PubMed]
7. Govindaraju, S.R.; Curry, B.D.; Bain, J.L.; Riley, D.A. Comparison of continuous and intermittent vibration effects on rat-tail artery and nerve. *Muscle Nerve* **2006**, *34*, 197–204. [CrossRef] [PubMed]
8. Kao, D.S.; Yan, J.G.; Zhang, L.L.; Kaplan, R.E.; Riley, D.A.; Matloub, H.S. Serological tests for diagnosis and staging of hand-arm vibration syndrome (HAVS). *Hand* **2008**, *3*, 129–134. [CrossRef] [PubMed]
9. Kanazuka, M.; Shigekiyo, T.; Toibana, N.; Saito, S. Increase in plasma thrombomodulin level in patients with vibration syndrome. *Thromb. Res.* **1996**, *82*, 51–56. [CrossRef] [PubMed]
10. Rossor, A.M.; Reilly, M.M. Blood biomarkers of peripheral neuropathy. *Acta Neurol. Scand.* **2022**, *146*, 325–331. [CrossRef] [PubMed]
11. Ising, E.; Åhrman, E.; Thomsen, N.O.B.; Eriksson, K.F.; Malmström, J.; Dahlin, L.B. Quantitative proteomic analysis of human peripheral nerves from subjects with type 2 diabetes. *Diabet. Med. J. Br. Diabet. Assoc.* **2021**, *38*, e14658. [CrossRef] [PubMed]
12. Notturno, F.; Capasso, M.; DeLauretis, A.; Carpo, M.; Uncini, A. Glial fibrillary acidic protein as a marker of axonal damage in chronic neuropathies. *Muscle Nerve* **2009**, *40*, 50–54. [CrossRef] [PubMed]
13. Tekavec, E.; Löfqvist, L.; Larsson, A.; Fisk, K.; Riddar, J.; Nilsson, T.; Nordander, C. Adverse health manifestations in the hands of vibration exposed carpenters—A cross sectional study. *J. Occup. Med. Toxicol.* **2021**, *16*, 16. [CrossRef] [PubMed]

14. Cooke, R.A.; Lawson, I.J. Cold intolerance and hand-arm vibration syndrome. *Occup. Med.* **2022**, *72*, 152–153. [CrossRef] [PubMed]
15. Carlsson, I.K.; Dahlin, L.B. Self-reported cold sensitivity in patients with traumatic hand injuries or hand-arm vibration syndrome an eight year follow up. *BMC Musculoskelet. Disord.* **2014**, *15*, 83. [CrossRef] [PubMed]

Proceeding Paper

Arterial Stenosis Stemming from Vibration-Altered Wall Shear Stress: A Way to Prevent Vibration-Induced Vascular Risk? [†]

Christophe Noël [1,*], **Maha Reda [1,2]**, **Nicla Settembre [3]** and **Emmanuelle Jacquet [2]**

[1] Electromagnetism, Vibration, Optics Laboratory, Institut national de recherche et de sécurité (INRS), 54519 Vandœuvre-lès-Nancy, France; maha.reda@inrs.fr

[2] Université de Franche-Comté, CNRS, Institut FEMTO-ST, F-25000 Besançon, France; emmanuelle.jacquet@univ-fcomte.fr

[3] Department of Vascular Surgery, Nancy University Hospital, University of Lorraine, 54500 Vandœuvre-lès-Nancy, France; nicla.settembre@univ-lorraine.fr

* Correspondence: christophe.noel@inrs.fr; Tel.: +33-383-502-112

† Presented at the 15th International Conference on Hand-Arm Vibration, Nancy, France, 6–9 June 2023.

Abstract: Vibration dose assessed by current standards is likely to be poorly suited to protecting workers against vibration white finger (VWF). Therefore, we intended for a two-step approach to better tackle vibration-induced pathophysiological vascular issues. In the first stage, a $\log_2$ linear regression law between the amplitude of vibration acceleration and the wall shear stress (WSS) drop was established. Then, in a second stage, we set up a mechanobiological model for computing the arterial stenosis stemming from the WSS decrease and encountered in patients suffering from VWF. Our findings highlighted a stenosis of about 30% when exposed for 10 years to a 40 m·s^{-2} amplitude vibration for 4 h a day.

Keywords: vibration white finger; wall shear stress; ultrasound; hyperplasia; arterial stenosis

Citation: Noël, C.; Reda, M.; Settembre, N.; Jacquet, E. Arterial Stenosis Stemming from Vibration-Altered Wall Shear Stress: A Way to Prevent Vibration-Induced Vascular Risk?. *Proceedings* **2023**, *86*, 11. https://doi.org/10.3390/proceedings2023086011

Academic Editor: Jacques Chatillon

Published: 10 April 2023

1. Introduction

Sustained exposure to high-level hand-transmitted vibrations may lead to angioneurotic disorders such as vibration-induced Raynaud's syndrome, also known as vibration white finger (VWF). Many physiological, histological, and epidemiological studies [1] have highlighted that the vibration dose assessed according to the current ISO 5349 standard may underestimate the onset predictions of such vascular injuries. In order to better tackle vibration-induced pathophysiological vascular issues, we recently set up a strategy in a two-step process [2]. First, we hypothesized that vibrations may acutely decrease the shear stress exerted by the blood flow on the artery endothelium layer. Second, studies have shown in various fields [3] that a chronic drop in this so-called wall shear stress (WSS) may result in arterial stenosis. Furthermore, angiographies and biopsies have emphasized this reduction in arterial lumen in patients suffering from VWF. Our approach then consisted of (i) assessing the relationship between vibration properties (frequency, amplitude) and the WSS drop and (ii) implementing a mechanobiological model which linked the vibration-reduced WSS and the resulting arterial stenosis. This current paper aims to establish how this strategy might pave the way for a new framework to prevent vibration-induced vascular risk.

2. Materials and Methods

2.1. Assessment of the Vibration-Altered Wall Shear Stress

An experimental device was set up to assess the vibration-induced WSS of the left proper volar forefinger artery at the level of the distal interphalangeal joint [4] while subjecting the right hand to mechanical vibration. The apparatus (Figure 1a) consisted mainly of an ultra-high-frequency ultrasound transducer connected to an ultrasound imaging system. In all, 24 volunteers in good health and who are nonsmokers, aged from 19 to 39 years old (average age 25.1), participated in a WSS measurement campaign in a room kept at a constant temperature ($23\,^{\circ}\text{C} \pm 0.5\,^{\circ}\text{C}$). The protocol consisted of estimating the WSS for three consecutive phases of 10 s each: (i) rest, (ii) exposure to vibrations, and (iii) return to calm. Vibration was a pure harmonic acceleration at 125 Hz for six amplitudes: 1, 2, 5, 10, 20, and 40 m·s^{-2} root mean square. WSS was then assessed by using a Womersley pulsatile flow model. The time averages of WSS (TAWSSs) over each of the three 10 s phases were subsequently worked out.

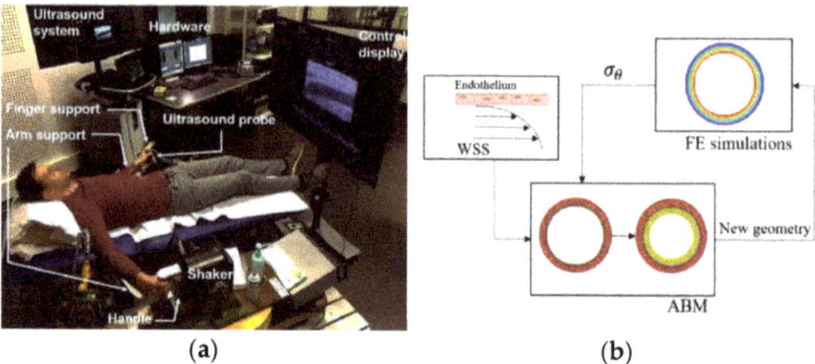

(a)

(b)

Figure 1. (**a**) Test setup for assessing the acute vibration-induced WSS; (**b**) The ABM-FE computational framework for simulating the chronic adaptive changes of the artery.

2.2. Modelling the Vibration-Induced Arterial Stenosis

A mechanobiological framework made up of an agent-based model (ABM) coupled with a finite elements (FE) model was set up (Figure 1b). The ABM caught up the hemodynamics-driven and mechanoregulated cellular and molecular mechanisms involved in vibration-induced intimal hyperplasia. Actually, this latter biological vascular process was assumed to likely be in part responsible for arterial stenosis. Many biological mechanisms were taken into account to model intimal hyperplasia, such as the secretion of mediators by the endothelial cells and the smooth muscle cells (SMCs), the proliferation/apoptosis and migration of SMCs, and the synthesis/degradation of extracellular matrix [5]. These phenomena were regulated by the WSS values, as well as the circumferential stresses (σ_θ) simulated by our FE model.

3. Result

3.1. Acute Impact of Vibrations on Arterial Hemodynamics

The drop in WSS triggered by the vibration occurred a few seconds after starting the vibratory excitation (Figure 2a).

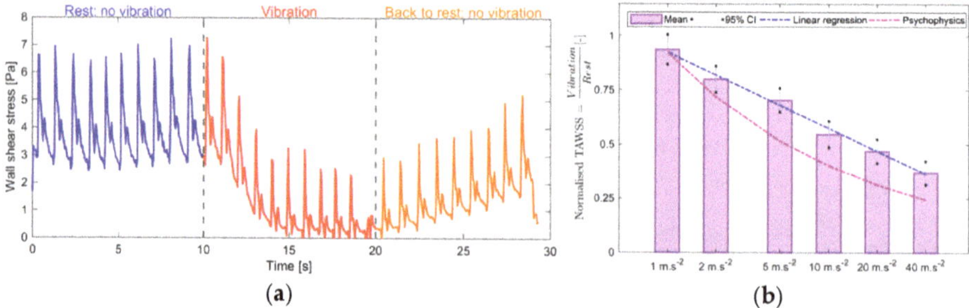

(a)

(b)

Figure 2. Effect of vibrations on the arterial blood flow: (**a**) WSS drop sparked by a vibration with an amplitude of 40 m·s^{-2} at 125 Hz; (**b**) statistical analysis of the influence of the vibration amplitude.

The TAWSSs normalized by the basal state (Figure 2b) depended on the amplitude of the vibration in a statistically significant way. They obeyed a $\log_2$ linear regression law of the vibration amplitude.

3.2. Chronic Response of the Artery at Tissue and Molecular Scales Due to Vibration-Induced WSS Drop

The normalized surface of the arterial lumen (Figure 3a) continually decreased with exposure time. The reduction in this surface was 12% at 5 years and 30% at 10 years of exposure. The mass per unit length of Matrix MetalloProtease MMP-2 (Figure 3b) accumulated continuously and considerably with the working lifetime, changing from the absence of MMP-2 initially to 50 pg/mm (picograms per mm) at the end of 10 years.

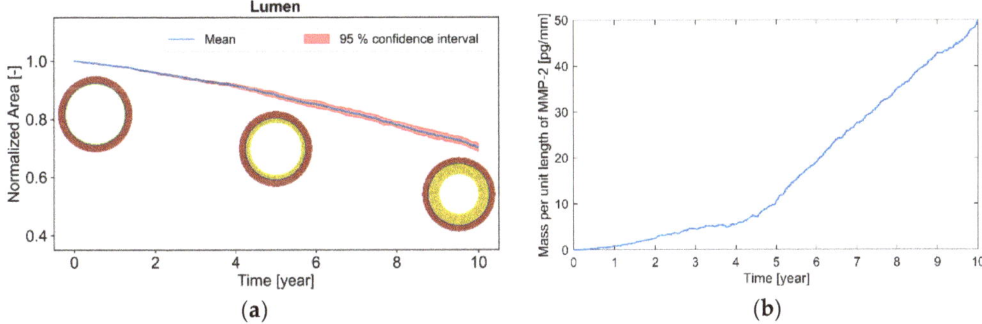

Figure 3. Chronic adaptive changes of the artery at the tissular and molecular scales: (**a**) normalized arterial lumen during the working life when exposed to vibrations for 4 h a day; (**b**) accumulation of MMP-2 during the working life.

3.3. Chart for Forecasting the Arterial Stenosis

The degree of arterial stenosis (expressed in %) is defined as the reduction in the arterial lumen normalized by its basal value. The abacus for forecasting the arterial stenosis (Figure 4) predicted a 20% arterial stenosis degree for an employee exposed to vibration for 2.5 h per day for 10 years. This same level of stenosis was also reached after 15 years of work for a daily exposure to vibrations of about 1 h 10 min.

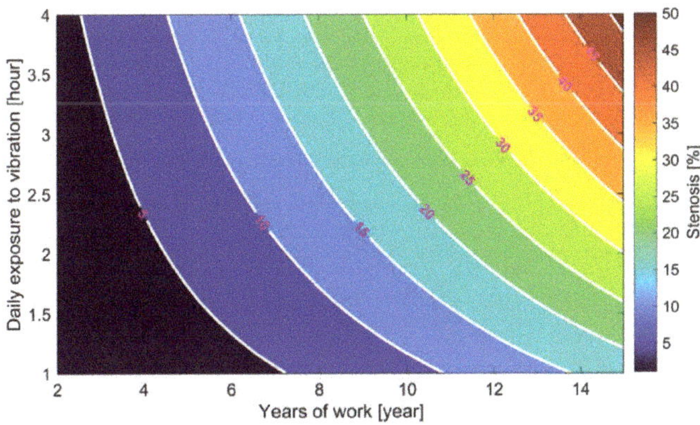

Figure 4. Degree of stenosis (%) computed according to the years of the working life and the daily exposure time to a vibration of an amplitude of 40 m·s^{-2} unweighted root mean square at 125 Hz.

4. Discussion

The development stage of VWF can be classified according to the degree of arterial stenosis: (i) type 0: healthy, (ii) disease type I: stenosis < 50%, (iii) disease type II: stenosis > 50%, (iv) disease type III: obstruction of the proper digital artery, and (v) disease type IV: obstruction of the common artery or upstream. Our model forecasted stenosis of around 30% after 10 years of exposure to a 40 m·s^{-2} amplitude vibration for 4 h a day, thereby leading to type I stenosis in keeping with the aforesaid classification.

Furthermore, a relationship between vibration amplitude and the subsequent WSS was established. Therefore, the WSS drop will be assessed by measuring in the field the vibration acceleration level on the handle of a vibrating machine. Thus, knowledge of the daily exposure (the model can take into account all types of exposure cycles) and the WSS (or similarly, the acceleration on the machine), the mechanobiological model will be permitted to work out the degree of stenosis and, thus, that of the disease for chronic exposure to vibrations.

Otherwise, with regard to the molecular upshots of chronic vibration exposure, MMP-2 accumulated substantially in the artery (Figure 3b). Thus, this enzyme could be a particularly suitable candidate biomarker for warning about and monitoring the evolution of VWF.

5. Conclusions

We succeeded in figuring out a relationship between acute vibration amplitude and the induced WSS drop. Our mechanobiological model was then able to forecast the chronic arterial stenosis elicited by that vibration-driven hemodynamic change. Linking vibration amplitude, daily exposure to vibration, working lifetime, and resulting stenosis is being used for building a new definition of vibration dose.

Author Contributions: Conceptualization, C.N. (all Sections), N.S. (Sections 2.1 and 3.1), M.R. (Sections 2.2 and 3.2), and E.J. (Sections 2.2 and 3.2); methodology, C.N. (all Sections), N.S. (Sections 2.1 and 3.1), M.R. (Sections 2.2 and 3.2), and E.J. (Sections 2.2 and 3.2); software, C.N. (all Sections) and M.R. (Sections 2.2 and 3.2); validation, C.N. (all Sections), N.S. (Sections 2.1 and 3.1), and E.J. (Sections 2.2 and 3.2); formal analysis, C.N. (all Sections), N.S. (Sections 2.1 and 3.1), M.R. (Sections 2.2 and 3.2), and E.J. (Sections 2.2 and 3.2); investigation, C.N.; resources, C.N.; data curation, C.N.; writing—original draft preparation, C.N.; writing—review and editing, C.N.; visualization, C.N.; supervision, C.N. (all Sections), N.S. (Sections 2.1 and 3.1), and E.J. (Sections 2.2 and 3.2); project administration, C.N. All authors have read and agreed to the published version of the manuscript.

Funding: This research received no external funding.

Institutional Review Board Statement: The study was conducted according to the guidelines of the Declaration of Helsinki and approved by the Institutional Review Board of the French National Agency for Medicines and Health Products (No. ID-RCB: 2018-A00614-651, 6 December 2018) and by the French National Ethical Research Committee (CPP 18 04 04, 6 December 2018).

Informed Consent Statement: Informed consent was obtained from all subjects involved in the study.

Data Availability Statement: B-mode cineloop and Pulsed Wave Doppler of the proper volar digital artery are available in Mendeley Data: https://data.mendeley.com/datasets/7g2p7t9tzt/1 (accessed on 10 January 2023).

Conflicts of Interest: The authors declare no conflict of interest.

References

1. *Mechanical Vibration—Measurement and Evaluation of Human Exposure to Hand Transmitted Vibration—Supplementary Method for Assessing Risk of Vascular Disorders*; ISO/TR 18570; ISO: Geneva, Switzerland, 2017.
2. Noël, C.; Settembre, N.; Reda, M.; Jacquet, E. A Multiscale Approach for Predicting Certain Effects of Hand-Transmitted Vibration on Finger Arteries. *Vibration* **2022**, *5*, 213–237. [CrossRef]
3. Humphrey, J.D. Vascular Adaptation and Mechanical Homeostasis at Tissue, Cellular, and Sub-Cellular Levels. *Cell Biochem. Biophys.* **2008**, *50*, 53–78. [CrossRef] [PubMed]

4.	Noël, C.; Settembre, N. Assessing Mechanical Vibration-Altered Wall Shear Stress in Digital Arteries. *J. Biomech.* **2022**, *131*, 110893. [CrossRef]
5.	Reda, M.; Noël, C.; Settembre, N.; Chambert, J.; Lejeune, A.; Rolin, G.; Jacquet, E. Agent-Based Model of the Vibration-Induced Intimal Hyperplasia. *Biomech. Model. Mechanobiol.* **2022**, *21*, 1457–1481. [CrossRef] [PubMed]

Proceeding Paper

Investigation of Hand-Arm Vibration (HAV) in Railroad Track Workers: Addressing Stakeholder Conflict of Interest [†]

Eckardt Johanning [1,*] and Paul Landsbergis [2]

1   Center for Family and Community Medicine, Columbia University and Johanning MD PC, New York, NY 10032, USA
2   School of Public Health, State University of New York (SUNY)-Downstate Health Sciences University, Brooklyn, NY 11203, USA
*   Correspondence: eckardtjohanning@johanningmd.com
†   Presented at the 15th International Conference on Hand-Arm Vibration, Nancy, France, 6–9 June 2023.

Abstract: The use of powered hand tools and equipment that exposes track workers to HAV vibration and other biomechanical hazards in the USA was investigated by a research team, following scientific principles and guidelines including protecting confidentiality of study participants. Musculoskeletal symptoms and neuro-musculoskeletal disorders were linked to workplace physical factors, such as HAV, and were reported in peer-reviewed journals. The methodology and results were subsequently challenged by a team of consultants hired by the Association of American Railroads (AAR), which represents major North American railroad corporations. Such an influence appears to challenge the integrity of occupational health research and impede the conduct of such research.

Keywords: physical hazard; hand-arm vibration; occupational epidemiology research; methods; conflict of interest; ethics; railroad; track workers; maintenance-of-way; epidemiology

1. Introduction

The National Institute of Safety and Health (NIOSH) estimated that more than 1.5–2 million workers in the United States (U.S.) are regularly exposed to hand-arm vibration (HAV) at work [1]. These old estimates did not include railroad maintenance-of-way (MoW) workers, who work on 140,000 miles of rails in the U.S., primarily owned by seven major class-one railroad corporations, represented by the Association of American Railroads (AAR). The job duties of a MoW worker have been often compared to general construction workers, but there are a number of specific and important differences regarding their special tools and work practices. Their work entails the use of heavy old-style hand tools and, more recently, makes use of powered hand tools and automated equipment that exposes workers to HAV. A university-based study team was tasked to collect epidemiological data from the members of the Brotherhood of Maintenance of Way Employees (BMWED) union. The results of the study were published recently in peer-reviewed journals [2–4]. The publications initially prompted an email inquiry to a co-investigator (P.L.) requesting access to our "raw data" because of "the overlap of interest", in order "to conduct additional analysis" by an "Occupational Epidemiologist" [5], without disclosure of the contractual affiliation of this "researcher" with the AAR and his other interested parties, who were hired by the AAR to reply to the published study results. The BMWED is the owner of the "raw data" and they denied this data request, indicating that the request had also been followed-up by additionally named "disclosed colleagues", of who all are well known for their robust "expert witness practices" on behalf of the major US railroads [6]. The union's concern was also based upon a "history of retaliation by railroad employers against employees who report injuries", a history which we documented in our reply [7]. This was followed up by individual letter(s) to the editors of the journals publishing our research by four named AAR consultants with academic and legal consulting firm affiliations.

These AAR consultants stated *"several concerning scientific limitations. Taken together, these limitations result in a manuscript that substantially misleads the readership on the associations between workplace factors and musculoskeletal disorders, as well as their conclusions on biomechanical risk for MOW employees"* [8–10]. The authors of these letters acknowledged that they received "partial funding" by the AAR for the preparation of the letters, but, other than methodological arguments, did not provide any compelling or objective data of their own in support of their assertions. The AAR consultants did cite conference proceedings of their AAR railroad-funded tool and equipment research; however, their research has never been published in peer-reviewed journals in order to allow the scientific community to evaluate their methods [11]. In detailed replies, the arguments of the AAR consults were addressed and a conflict of interest (CoI) of these authors was raised [7,11,12]. The challenge for occupational health research is to avoid or control any CoI by the funding sources or other influences.

2. Original Study Results

Briefly, the details of the methods, materials and key findings of our MoW study were reported in the three peer-reviewed journals based on a comprehensive questionnaire addressing work practices, work factors and the health of approximately 34,580 current BMWED members and 3975 retirees. Survey responses were received between 1 August 2016 and 28 February 2017. The survey was answered by 4816 members and retirees in full or in part, amounting to approximately a 12% response rate. The survey questions on vibration exposure and symptoms were based on validated instruments, primarily the collaborative European VIBRISKS project [13]. Musculoskeletal symptoms and disorders were reported by the MoW survey respondents [2–4].

MoW workers frequently reported typical hand-transmitted vibration-related symptoms, and appeared to be at risk for neuro-musculoskeletal disorders of the upper extremity. Compared to all U.S. employed men aged 18–74 years, active BMWED men were more likely to have been told by a doctor or a health professional that they have carpal tunnel syndrome (CTS) (7.9% vs. 3.6%) [2]. Daily or weekly symptoms during the past year, consistent with vibration-related disease (fingers going white (blanching) when exposed to cold (n = 143, 3.7%)), and having experienced white fingers where the whiteness was clearly demarcated (showed clear limits or boundaries) (n = 77, 2.0%). In addition, 8.0% (n = 314) reported difficulty picking up very small objects, such as screws or buttons, or opening tight jars [4]. In addition, workers reported biomechanical, WBV and HAV exposures, and associations between those exposures and health outcomes.

Compared with U.S. employed men, adjusted for age, race and region, active male MOW workers were more likely to report "repeated lifting, pushing, pulling, or bending" at work (74.6% vs. 46.9%), and not enough staff (88.1% vs. 65.2%). They were less likely to report management priority on workplace health and safety (59.37% vs. 94.8%), ability to make job decisions on their own (68.4% vs. 87.7%), and supervisor support (60.3% vs. 90.8%) (all comparisons, p < 0.001) [2].

Associations were found between the use of high-vibration vehicles and neck pain (aPR = 1.47, 95% confidence interval (CI): 1.07–2.03) and knee pain (aPR = 1.38, 95% CI: 1.04–1.82) for more than 1.9 years (vs. 0 year) of full-time equivalent use, but not back pain. Back pain radiating below the knee (sciatica indicator) was associated with high-vibration vehicle use greater than 0.4 and less than 1.9 years (aPR = 1.58, 95% CI: 1.15–2.18) [3].

Powered hand tools were ranked according to each tool's listed average segmental vibration emissions and self-reported average use. In the analyses of specific tool-related work exposures and shoulder, elbow and hand/wrist symptoms, the ranked frequency of tool use, as reported by the MoW worker survey participants, was calculated (adjusted for age, region, race/ethnicity, smoking, second job vehicle vibration, spare time vehicle vibration) and listed in two detailed tables of the article. The average duration of full time equivalent exposure values ranged from 5.04 y of vibration exposure (impact wrench) to 0.06 y of vibration exposure (scabbler). It is noteworthy that 50% of users of nine of the ten highest ranked tools in the tables indicated that they "always" or "often" used that tool. A significantly increased risk in pain was seen after 10 y (x fraction of a day) use of various powered hand tools, ranging from 32% (asphalt tamper, $n = 336$) to 71% (nut splitter, $n = 176$) increased risk [4]. A literature review and analysis of HAV emissions of MoW hand tools was conducted using published resources by independent, governmental and commercial/manufacturer sources [4]. Of all of the powered hand tools used by this track worker trade, 88% of the selected tools exceeded a 5 m/s^2 emission level and were above vibration emission magnitudes of common tools of other comparable industries.

3. Summary and Conclusions

Scientific publications about identified workplace hazards that require employer attention may have legal consequences under the US Federal Employers Liability Act (FELA), as a "notice to employers", i.e., the US railroad corporations, showing "negligence and proximate cause". In our replies to the Letter(s)-to-the-Editor by the AAR consultants, we emphasized that we followed guidance from our respective Institutional Review Boards (IRBs) and an international and independent scientific advisory panel regarding assuring participant confidentiality and protection from reprisals, and the methodology, design of the survey instrument, and data acquisition and analysis. Potential study limitations, including the use of workers' self-report of symptoms and workplace exposures, and a low survey response rate, were acknowledged, and the methods for statistical adjustments were described in the replies to the letters [7,11,12].

Corporate influence on public and occupational health research, challenges to research study reports and publication of biased science has a long history in the United States [14–16]. Such influence is designed to challenge the credibility of occupational health research. We fully support the idea that scientific integrity is based on the principle that research is conducted as objectively as possible. Public health and occupational health rely on the integrity, objectivity and validity of data, as well as disclosing all possible CoIs of the authors. Physician-industry or -Union collaboration may trigger the "imputation of motive" and a raised "doubt" or "concern" is meant to be an assault on science [14–16]. Occupational safety and health research needs to recognize and manage these possible CoIs when investigating workplace conditions, and investigators should "design and carry out their activities on a sound scientific basis with full professional independence and follow the ethical principles relevant to health and medical research work" [17]. CoIs can be controlled by adhering to a professional "Code of Ethics", such as that outlined by the American College of Occupational and Environmental Medicine (ACOEM) or the Code of Ethics of the International Commission on Occupational Health (ICOH). The primary responsibility should be the health and safety of the individual in the workplace and the environment [18]. A cooperative study by the stakeholders could occur if the AAR and the railroad companies accept the BMWED invitation for a joint study. Researchers from NIOSH have offered technical assistance for any further follow-up intervention studies. Such a study could include a review of anonymous medical claim and disability data specific to recognized HAV related medical conditions. Such studies could also include evaluations of interventions involving ergonomically designed hand tools and equipment with improved vibration attenuation technology.

Author Contributions: Conceptualization, E.J.; writing—original draft preparation, E.J.; writing—review and editing, E.J. and P.L. All authors have read and agreed to the published version of the manuscript.

Funding: This research paper writing, review and editing received no external funding. The cited study was funded by a contract from the Brotherhood of Maintenance of Way Employees Division (BMWED), International Brotherhood of Teamsters.

Institutional Review Board Statement: The study [2–4] was conducted in accordance with the Declaration of Helsinki, and approved by the Institutional Review Board of the Cook County Hospital (Chicago, Illinois) and the State University of New York-Downstate Medical Center (Brooklyn, New York). To ensure that the identity of all survey participants would be legally protected from discovery, a Certificate of Confidentiality was issued by the National Institutes of Health (NIH), USA.

Informed Consent Statement: Informed consent was obtained from all subjects involved in the study [2–4].

Data Availability Statement: No new data were created or analyzed in this study. Data sharing is not applicable to this article. Cited study data of the referenced publications is under the ownership of the BMWED.

Conflicts of Interest: Johanning has been evaluating and treating railroad workers and represented some in Federal Employers Liability Act (FELA) disability claims. Landsbergis is a consultant to the Center for Social Epidemiology, Marina Del Rey, CA, on issues related to work organization. The BMWED funded the MoW study and facilitated access to the union's membership for informed consent and survey administration, but had no role in the analyses, or interpretation of data; in the writing of the manuscripts; or in the decision to publish the results.

References

1. The National Institute for Occupational Safety and Health (NIOSH). Vibration Syndrome: Current Intelligence Bulletin 38. 1983. Available online: https://www.cdc.gov/niosh/docs/83-110/default.html (accessed on 4 January 2023).
2. Landsbergis, P.; Johanning, E.; Stillo, M.; Jain, R.; Davis, M. Work Exposures and Musculoskeletal Disorders Among Railroad Maintenance-of-Way Workers. *J. Occup. Environ. Med.* **2019**, *61*, 584–596. [CrossRef] [PubMed]
3. Landsbergis, P.; Johanning, E.; Stillo, M.; Jain, R.; Davis, M. Upper extremity musculoskeletal disorders and work exposures among railroad maintenance-of-way workers. *Am. J. Ind. Med.* **2021**, *64*, 744–757. [CrossRef] [PubMed]
4. Johanning, E.; Stillo, M.; Landsbergis, P. Powered-hand tools and vibration-related disorders in US-railway maintenance-of-way workers. *Ind. Health* **2020**, *58*, 539–553. [CrossRef]
5. Thiese, M.S. (University of Utah School of Medicine) E-mail message to author Landsbergis, 4 March 2020.
6. Voegel, Z. (BMWED) 2020. E-mail reply to Thiese request for raw data. Unpublished. Pers. communication: (available upon request).
7. Landsbergis, P.; Johanning, E.; Stillo, M. Landsbergis, Johanning, Stillo Respond to Letter to the Editor. *J. Occup. Environ. Med.* **2021**, *63*, e751–e754. [CrossRef]
8. Thiese, M.S.; Hegmann, K.T.; Page, G.B.; Weames, G.G. Letter to the Editor: Landsbergis et al. (2019) Titled "Work Exposures and Musculoskeletal Disorders Among Railroad Maintenance-of-Way Workers". *J. Occup. Environ. Med.* **2021**, *63*, e745–e750. [CrossRef] [PubMed]
9. Thiese, M.S.; Hegmann, K.T.; Weames, G.G.; Page, G.B. Concerns re Landsbergis et al.: Occupational risk factors for musculoskeletal disorders among railroad maintenance-of-way workers. *Am. J. Ind. Med.* **2021**, *64*, 714–716. [CrossRef] [PubMed]
10. Weames, G.G.; Page, G.B.; Thiese, M.S.; Hegmann, K.T. Concerns regarding the publication "Powered-hand tools and vibration-related disorders in US-railway maintenance-of-way workers". *Ind. Health* **2022**, *60*, 284–287. [CrossRef] [PubMed]
11. Landsbergis, P.; Johanning, E.; Stillo, M. Landsbergis et al. respond. *Am. J. Ind. Med.* **2021**, *64*, 717–720. [CrossRef] [PubMed]
12. Johanning, E.; Stillo, M.; Landsbergis, P. Reply to "raised concern". *Ind. Health* **2022**, *60*, 288–292. [CrossRef] [PubMed]
13. European Commission. Risks of Occupational Vibration Exposures Vibrisks—FP5 Project No. QLK4-2002-02650. 2007. Available online: http://www.vibrisks.soton.ac.uk/reports/Annex21%20UTRS_AMC%20WP1_3_%204_3%20070307.pdf (accessed on 4 January 2023).
14. Okike, K.; Kocher, M.S.; Mehlman, C.T.; Bhandari, M. Industry-sponsored research. *Injury* **2008**, *39*, 666–680. [CrossRef] [PubMed]
15. Baur, X.; Soskolne, C.L.; Bero, L.A. How can the integrity of occupational and environmental health research be maintained in the presence of conflicting interests? *Environ. Health* **2019**, *18*, 93. [CrossRef] [PubMed]
16. Michaels, D. *Doubt Is Their Product: How Industry's Assault on Science Threatens Your Health*; Oxford University Press: New York, NY, USA, 2008.
17. International Commission on Occupational Health (ICOH). International Code of Ethics: For Occupational Health Professionals. 2014. Available online: https://www.icohweb.org/site/code-of-ethics.asp (accessed on 4 January 2023).
18. ACOEM Code of Ethics. American College of Occupational and Environmental Medicine. Available online: https://acoem.org/about-ACOEM/Governance/Code-of-Ethics (accessed on 4 January 2023).

Abstract

Onset of Vibration-Induced White Finger: Insight Derived from a Meta-Analysis of Exposed Workers [†]

Magdalena F. Scholz [1,*], Anthony J. Brammer [2] and Steffen Marburg [1]

[1]   Chair of Vibroacoustics of Vehicles and Machines, Technical University of Munich, 85748 Garching, Germany
[2]   Department of Medicine, University of Connecticut Health, Farmington, CT 06030, USA; brammer@uchc.edu
[*]   Correspondence: magdalena.scholz@tum.de; Tel.: +49-89-289-55133
[†]   Presented at the 15th International Conference on Hand-Arm Vibration, Nancy, France, 6–9 June 2023.

Abstract: A pooled analysis has been performed of population groups whose hands have been occupationally exposed to vibration to study exposure evaluation, using epidemiologic data selected from a published meta-analysis. While the analysis cannot confirm the accuracy of the exposure-response relation in ISO 5349-1:2001, it suggests that the relation provides a conservative estimate for the onset of vibration-induced white finger. The analysis also demonstrates that the procedures for calculating vibration exposure in the international standard may need revision.

Keywords: hand-arm vibration; exposure-response relation; prevalence; ISO 5349-1:2001; prevalence prediction model

The onset of vibration-induced white finger (VWF) in workers operating power tools or machines from which vibration enters the hands is a subject of considerable interest for establishing occupational health exposure limits. Guidelines have been proposed from epidemiologic studies and incorporated into regulations and standards. A continuing debate has focused on the accuracy of the guidelines in the international standard for hand-transmitted vibration, ISO 5349-1:2001. A comprehensive meta-analysis of studies on the health effects of workers whose hands have been exposed to vibration has recently been conducted by Nilsson et al. [1]. The data gathered from the studies are further employed to create a model to predict a 10% prevalence of VWF in a population group occupationally exposed to vibration.

In this contribution, described in Ref. [2], relations between the mean lifetime duration of exposure (in years), D_y, and the vibration entering the hands, expressed by the daily 8 h, energy-equivalent, frequency-weighted acceleration, A(8), are constructed from studies ranked acceptable by Nilsson et al. [1], and compared with the ISO predictions. For this purpose, additional rules are introduced to confirm: (1) compliance with the measurement procedures in ISO 5349-1:2001, and (2) that the signs and symptoms were most likely caused by vibration exposure. Data sets have been formed first for studies reporting D_y and A(8) (including values derived from vibration spectra), and second, for studies in which D_y has been reconstructed from hourly exposures. The reported point prevalences and mean group lifetime exposures are employed to estimate, by linear interpolation, the times at which 10% of the groups of workers are estimated to have been affected by VWF. This estimated time is plotted against the A(8) value for every group of workers. Models are created by the means of regression analyses of these resulting data sets, assuming the same form of relation as that described in ISO 5349-1:2001.

Limiting the analyses to data that allow for interpolation, and therefore excluding studies that would have required extrapolation results in models with 95-percentile confidence intervals that include the ISO model predicting 10% prevalence of VWF. Furthermore, differences can be observed between models created from studies in which the workers used only one power tool per day and those obtained for studies in which workers experience a daily exposure to multiple tools and machines. Within the analyses, it is also observed that very different prevalences are recorded in studies with comparable A(8) values and lifetime exposures. Both of these observations suggest that the procedure for calculating daily exposure in ISO 5349-1:2001 may need revision. The analyses and their results are shown in detail in Ref. [2].

In summary, while the analyses described in Ref. [2] cannot confirm either the validity of the exposure-response relation in ISO 5349-1:2001 or the need for its revision, it appears to form a conservative estimate of the onset of VWF in a population group exposed to hand-transmitted vibration. Further analyses are needed in order to determine the factors influencing the daily exposure found in Ref. [2], in particular, the calculation of exposure when multiple tools or machines are used during a workday, and the formulation of the magnitude of an exposure. Additionally, more recent studies and those in languages other than English presently omitted from the meta-analysis of Nilsson et al. [1] will need to be considered.

Author Contributions: Conceptualization, A.J.B. and M.F.S.; data acquisition and statistical analysis, M.F.S.; writing—review and editing, M.F.S. and A.J.B.; supervision, S.M. All authors have read and agreed to the published version of the manuscript.

Funding: This research received no external funding.

Institutional Review Board Statement: Not applicable.

Informed Consent Statement: Not applicable.

Data Availability Statement: All results and data can be found in Ref. [2].

Conflicts of Interest: The authors declare no conflict of interest.

References

1. Nilsson, T.; Wahlström, J.; Burström, L. Hand-arm vibration and the risk of vascular and neurological diseases—A systematic review and meta-analysis. *PLoS ONE* **2017**, *12*, e0180795. [CrossRef] [PubMed]
2. Scholz, M.F.; Brammer, A.J.; Marburg, S. Exposure-response relation for vibration-induced white finger: Inferences from a published meta-analysis of population groups. *Int. Arch. Occup. Environ. Health* **2023**. [CrossRef] [PubMed]

Proceeding Paper

Dose-Response Relationship between Hand-Arm Vibration Exposure and Musculoskeletal Disorders of Upper Extremities: A Case-Control Study among German Workers [†]

Yi Sun [1,*], Frank Bochmann [1], Winfried Eckert [2], Benjamin Ernst [1], Uwe Kaulbars [1], Uwe Nigmann [3], Nastaran Raffler [1], Christina Samel [1] and Christian van den Berg [4]

[1] Institute for Occupational Safety and Health of the German Social Accident Insurance (IFA), 53757 Sankt Augustin, Germany; frank.bochmann@dguv.de (F.B.); benjamin.ernst@dguv.de (B.E.); uwe.kaulbars@dguv.de (U.K.); nastaran.raffler@dguv.de (N.R.); christina.samel@dguv.de (C.S.)

[2] Institution for the Building Trade, German Social Accident Insurance, 71032 Boeblingen, Germany; winfried.eckert@bgbau.de

[3] Institution for the Woodworking and Metalworking Industries, German Social Accident Insurance, 40472 Dusseldorf, Germany; uwe.nigmann@bghm.de

[4] Institution for the Raw Materials and Chemical Industry, German Social Accident Insurance, 44789 Bochum, Germany; christian.vandenberg@bgrci.de

* Correspondence: yi.sun@dguv.de; Tel.: +49-30-130013123

† Presented at the 15th International Conference on Hand-arm Vibration, Nancy, France, 6–9 June 2023.

Abstract: In an epidemiological case-control study, exposure-response relationship between hand-arm vibration exposure and the risk of musculoskeletal disorders (MSDs) of the upper extremities were examined among 209 male cases and 614 controls in the German construction, mining, metal, and wood-working industries. To quantify individual vibration exposures, a database of industrial hygiene measurements of over 700 power tools was established. In addition, individual work histories were collected in detail. The dose-response relationships between hand-arm vibration exposure and the risk of MSDs were quantified based on multivariable logistic regression analysis. After adjusting for relevant confounders, statistically significant dose-response relationships between cumulative hand-arm vibration exposure doses and MSDs of the upper extremities could be established.

Keywords: hand-arm vibration; musculoskeletal disorders; dose-response relationship; epidemiology; risk assessment

Citation: Sun, Y.; Bochmann, F.; Eckert, W.; Ernst, B.; Kaulbars, U.; Nigmann, U.; Raffler, N.; Samel, C.; van den Berg, C. Dose-Response Relationship between Hand-Arm Vibration Exposure and Musculoskeletal Disorders of Upper Extremities: A Case-Control Study among German Workers. *Proceedings* **2023**, *86*, 22. https://doi.org/10.3390/proceedings2023086022

Academic Editors: Christophe Noël and Jacques Chatillon

Published: 12 April 2023

1. Introduction

Health related effects of hand-arm vibration exposure have three main clinical components: vascular, neurological, and musculoskeletal disorders [1]. Vascular and neurological disorders usually occur together and are the most extensively studied forms of hand-arm vibration syndrome [1]. In contrast, vibration induced musculoskeletal disorders (MSDs) of the upper extremities have been less extensively studied. Early studies and reviews indicate elevated risk of musculoskeletal symptoms (upper limb pain, stiffness, and muscle tendon syndrome) and osteoarthritis (OA) among vibration-exposed workers [1–4]. However, the exposure-response relationship between hand-arm vibration exposure and MSDs of upper extremities has been poorly established to date.

To quantitatively evaluate the exposure-response relationship between hand-arm vibration exposure and the risk of MSDs, an epidemiological case-control study was carried out among workers in the construction, mining, metal, and wood-working industries in Germany.

2. Material and Methods

2.1. Design and Study Population

In a multicentre industry-based case-control study, 209 consecutive male cases and 614 controls were recruited during the time between 2010 and 2021. Cases were newly reported patients with the following six clinical outcomes: hand OA, elbow OA, shoulder OA, Kienböck's disease, Elbow Osteochondrosis, Scaphoid fracture, and scaphoid pseudoarthrosis. Controls were a random sample of persons with compensable occupational injuries resulting in at least 3 days away from their job. They were matched to the cases in a ratio of about 1:3 for birth year, industrial sectors, and reporting years. Standardized personal interviews were carried out among cases and controls by well-trained safety engineers. In addition to leisure activities and comorbidities, work histories of all participants were collected in detail.

2.2. Exposure Assessment

To quantify hand-arm vibration exposures, a database of industrial hygiene measurements of over 700 power tools was established. This database allows for detailed quantification of vibration exposures over time. The cumulative vibration doses are quantified as the sum-of-squares of daily vibration exposure over the period of whole working life:

$$\begin{aligned} D_{hv} &= \sum a_{hvi(8)}^2 d_i \\ D_{hw} &= \sum a_{hwi(8)}^2 d_i \end{aligned} \tag{1}$$

where

D_{hv} = cumulative vibration doses in three measuring directions;
D_{hw} = cumulative vibration doses in the direction along the forearm;
$a_{hvi(8)}$ = daily vibration exposure of three measuring directions at day i;
$a_{hwi(8)}$ = daily vibration exposure in the direction along the forearm at day i;
d_i = number of working days with a daily exposure of $a_{hvi(8)}$ ($a_{hwi(8)}$).

2.3. Statistical Analysis

The dose-response relationship between hand-arm vibration exposure and MSDs of the upper extremities was quantified by conditional logistic regression analyses adjusting for age, sex, study centre, generalized OA, injury, and inflammatory disorder of hand, elbow, and shoulder joints.

3. Results

The study sample (n = 823) has an average age of about 52 (range: 22–84) years, and an average employment duration of around 24 (range: 0.49–49.43) years. The individual work history contains a total of 5115 exposure sections over an exposure period of about 50 years. A total of 423 technical power tools were identified which induce hand-arm vibration exposures.

Cases have, on average, about 48% more cumulative working hours with hand-held technical power tools than the controls. This leads to higher cumulative vibration exposure doses among the cases than among the controls (Figure 1).

Table 1 gives the estimated exposure-response relationships between cumulative vibration exposure (D_{hv} and D_{hw}) and the risk of musculoskeletal disorders of the upper extremities. The statistical models demonstrate consistent and significant exposure-response relationships between cumulative hand-arm vibration exposure (D_{hv} and D_{hw}) and the risk of musculoskeletal disorders. Due to the strong correlation of the D_{hv} and D_{hw}, there are similar effect sizes in exposure-response relationships for the D_{hv} and D_{hw} (s. Table 1).

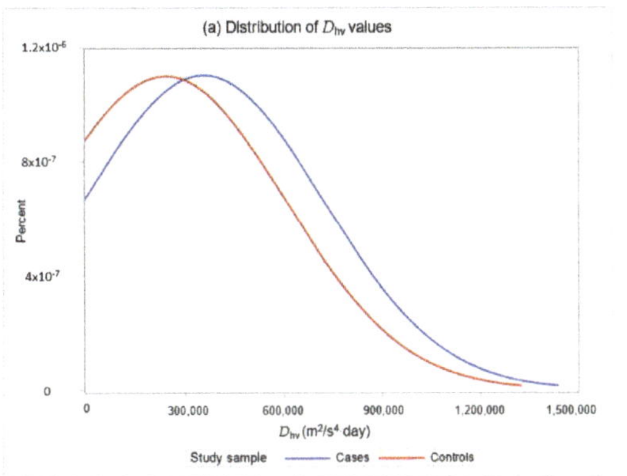
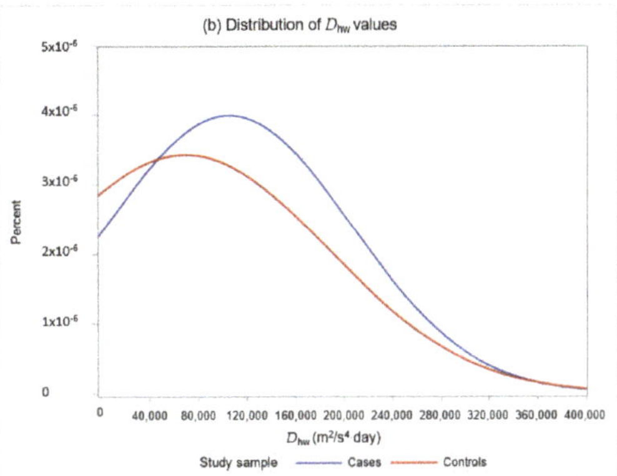

Figure 1. Distribution of cumulative vibration doses (D_{hv} and D_{hw}) among cases and controls ($n = 823$).

Table 1. Dose-response relationship between cumulative vibration exposure (D_{hv}, D_{hw}) and musculoskeletal disorders.

	Cases/N	Unadjusted		Adjusted *	
		OR	95%CI	OR	95%CI
D_{hv} (m^2/s^4 · day)					
1. Quintile	20/165	1	–	1	–
2. Quintile	35/164	2.14	1.17–3.90	2.08	1.12–3.85
3. Quintile	46/165	3.10	1.72–5.59	2.66	1.45–4.88
4. Quintile	40/164	2.77	1.52–5.06	3.31	1.78–6.13
5. Quintile	68/165	5.03	2.83–8.93	5.65	3.06–10.42
Trend-test		$p < 0.0001$		$p < 0.001$	
100 m^2/s^4 · year increase		1.015	1.008–1.023	1.013	1.006–1.021
D_{hw} (m^2/s^4 · day)					
1. Quintile	16/165	1	–	1	–
2. Quintile	27/164	1.93	1.02–3.67	1.73	0.89–3.33
3. Quintile	44/165	3.57	1.92–6.62	3.19	1.70–6.01
4. Quintile	58/164	4.91	2.68–8.99	3.92	2.10–7.32
5. Quintile	64/165	5.08	2.80–9.22	4.43	2.39–8.21
Trend-test		$p < 0.0001$		$p < 0.0001$	
100 m^2/s^4 · year increase		1.036	1.015–1.058	1.028	1.006–1.050

* Adjusted for study centers, generalized OA, injuries, and inflammatory disorders of hand, elbow, and shoulder joints.

Based on the effect estimates given in Table 1, smooth lines of exposure-response curves were quantified with the conventional log-linear exposure-response assumptions (s. Figure 2). According to the estimated smooth lines of dose-response curves, a 10%, 30%, and doubled increased risk of musculoskeletal disorders can be expected, as shown in Figure 3. A combination of working days and expected daily vibration exposure gives the exact vibration doses that can lead to the expected excess risk of musculoskeletal disorders (s. Figure 3).

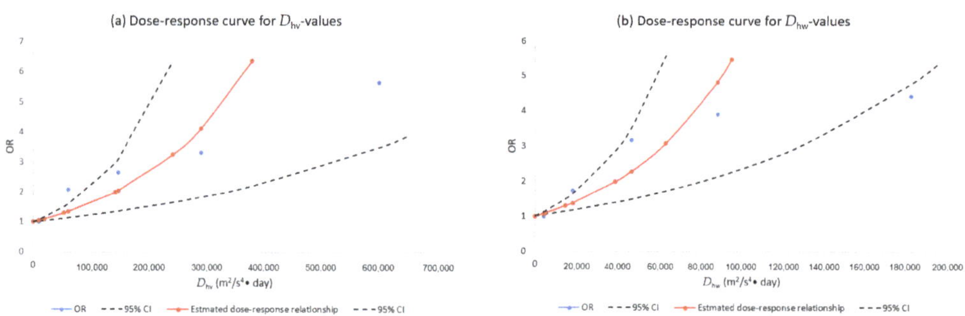

Figure 2. Estimated exposure-response curves for D_{hv} and D_{hw} values.

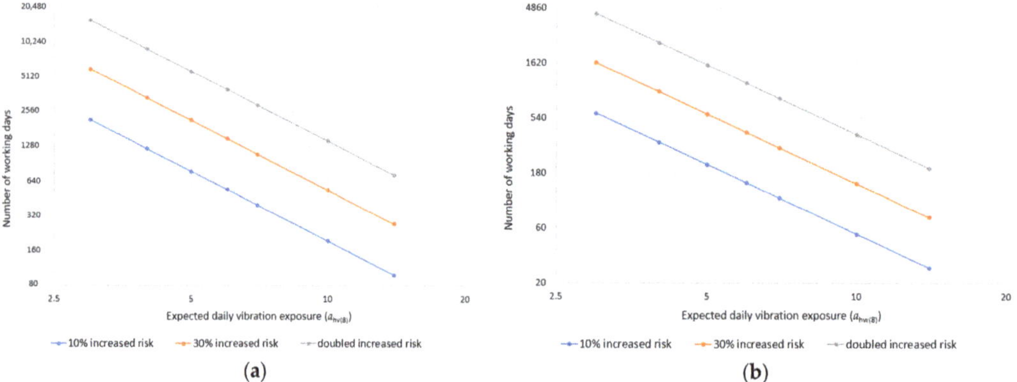

Figure 3. Expected 10%, 30%, and doubled increased risk of musculoskeletal disorders. (**a**) Dose-response-relationship for daily vibration exposure of three measuring directions ($a_{hv(8)}$). (**b**) Dose-response-relationship for daily vibration exposure in the direction along the forearm ($a_{hw(8)}$).

4. Conclusions

Overall, this study has a large sample size and high methodological quality and shows for the first time an exposure-response-relationship between hand-arm vibration exposure and the risk of musculoskeletal disorders of the upper limb based on clearly defined morphological changes. The findings of this study provide useful guidance in the prevention and compensation of work-related and vibration-induced musculoskeletal disorders of the upper limbs.

Author Contributions: Conceptualization, Y.S., F.B. and U.K.; methodology, Y.S., F.B. and U.K.; formal analysis, Y.S.; investigation, W.E., U.N. and C.v.d.B.; data curation, Y.S., U.K., N.R. and B.E.; writing—original draft preparation, Y.S.; writing—review and editing, Y.S., F.B., U.K., N.R., B.E. and C.S.; project administration, F.B. and W.E.; funding acquisition, W.E. All authors have read and agreed to the published version of the manuscript.

Funding: This research was funded by the German Social Accident Insurance, grant number FP-297.

Institutional Review Board Statement: This study was conducted in accordance with the Declaration of Helsinki, and approved by the Institutional Review Board of the German Social Accident Insurance (protocol code FP-297, 15 September 2009).

Informed Consent Statement: Informed consent was obtained from all subjects involved in the study.

Data Availability Statement: Due to the nature of this research, participants of this study did not agree for their data to be shared publicly, so supporting data is not available.

Conflicts of Interest: The authors declare no conflict of interest.

References

1. House, R.; Wills, M.; Liss, G.; Switzer-McIntyre, S.; Manno, M.; Lander, L. Upper extremity disability in workers with hand-arm vibration syndrome. *Occup. Med.* **2009**, *59*, 167–173. [CrossRef] [PubMed]
2. Hagberg, M. Clinical assessment of musculoskeletal disorders in workers exposed to hand-arm vibration. *Int. Arch. Occup. Environ. Health* **2002**, *75*, 97–105. [CrossRef] [PubMed]
3. Bovenzi, M. Health risks from occupational exposures to mechanical vibration. *Med. Lav.* **2006**, *97*, 535–541. [PubMed]
4. Charles, L.E.; Ma, C.C.; Burchfiel, C.M.; Dong, R.G. Vibration and ergonomic exposures associated with musculoskeletal disorders of the shoulder and neck. *Saf. Health Work* **2018**, *9*, 125–132. [CrossRef] [PubMed]

Proceeding Paper

Hand-Arm Vibration Syndrome in Dentistry: A Questionnaire Survey among Dentists and Review of Literature [†]

Alice Turcot *, Denis Hamel and Mélanie Tessier

Institut National de Santé Publique du Québec, Quebec City, QC G1V 5B3, Canada;
denis.hamel@inspq.qc.ca (D.H.); melanie.tessier@inspq.qc.ca (M.T.)
* Correspondence: alice.turcot.med@ssss.gouv.qc.ca
† Presented at the 15th International Conference on Hand-Arm Vibration, Nancy, France, 6–9 June 2023.

Abstract: The use of dental handpieces and ultrasonic instruments expose dental professionals to high-frequency vibration, precise gripping, high pinch force, and repetitive bending movements of wrist during restorative procedures involving cutting dental material, periodontal scaling, and root planning. There is clear evidence of an association between the dentistry profession and work-related musculoskeletal disorders in the neck, upper back and upper extremities; however, the influence of high-frequency vibration on hand and fingers from dental handpieces is not well known. The objectives of the current paper are to present the results of a survey on hand-arm vibration syndrome (HAVS) among members of a professional dental society and to present a literature review on dental handpieces and ultrasonic scalers exposure assessment and occurrence of hand-arm vibration syndrome among dental professionals. There seems to be limited awareness of the occupational risk associated with hand-arm vibration from handpieces and ultrasonic devices. This study highlights the occurrence of vascular and neurological disorders of HAVS among dental professionals, as well as wrist/hand pain, osteoarthritis, diminished hand grip, and carpal tunnel syndrome. The assessment of high-frequency vibration and ultra-vibration from these vibrating tools and vibration-related injuries deserve special attention for future preventive measures.

Keywords: dentists; musculoskeletal disorders; dental handpieces; hand-arm vibration syndrome; white fingers; neuro-sensorineural injury; exposure; vibration; wrist/hand/fingers pain; carpal tunnel syndrome; osteoarthritis

Citation: Turcot, A.; Hamel, D.; Tessier, M. Hand-Arm Vibration Syndrome in Dentistry: A Questionnaire Survey among Dentists and Review of Literature. *Proceedings* **2023**, *86*, 17. https://doi.org/10.3390/proceedings2023086017

Academic Editors: Christophe Noël and Jacques Chatillon

Published: 11 April 2023

1. Introduction

Handpieces and ultrasonic scalers expose dental professionals, including dental specialists and dental hygienists, to high-frequency mechanical vibration, ranging from 0.5 kHz to 50 kHz. Air turbines and micromotor handpieces are used for tooth preparation, removal of decay, root canal treatments, restorations, implants surgery, bone cutting procedures, and various other procedures. Traditional handpieces are either air-driven or electrically driven, running with low or high speed. Each tool has specific purposes for dental procedures [1]. Dentists and dental hygienists use low-speed micromotor handpieces for polishing and removing decays, which are typically used between 20,000 rpm and 40,000 rpm. High-speed electric handpieces have typical speeds in the range of 200,000 rpm, while high speed air-driven handpieces operate at up to 400,000 rpm, but are typically and are usually used within the 180,000 to 330,000 rpm range. In addition to handpieces, sonic and ultrasonic scalers are frequently used by dentists, specialists, and dental hygienists for periodontal procedures [2]. The latter operate in a wide range of frequencies: around 3–8 kHz for sonic scalers, 18–45 kHz for piezoelectric ultrasonic scalers, and 25–50 kHz for magnetostrictive scalers [3]. Early studies have shown that dental professionals exposed to high-frequency dental tools have shown neurological and vascular symptoms, especially in the dominant hand, comparable to hand-arm vibration syndrome dating back to 1980 [4,5].

The pathological role of high-frequency vibration and potential risk for vibration-related disorders among dental professionals remains to be fully understood. Despite the fact that these excitations are much higher that the hand natural frequencies, it was postulated that injury to small nerve fibers and mecanoreceptors in the skin would be related to vibration neuropathic disorder. In addition, high-frequency vibration may contribute to decreased hand grip, pain, carpal tunnel syndrome, and osteoarthritis, although typically associated with working position, repetitive movements, and high biomechanical load. To date, no study on HAVS among dentists in Quebec has been undertaken. This study presents the results of a short questionnaire survey among dentists and a narrative review of literature to estimate the prevalence of HAVS and musculoskeletal wrist/hand pain, osteoarthritis, and carpal tunnel syndrome among dental professionals exposed to vibrating dental tools.

2. Methods

A preliminary survey was carried out on dentists from a dental association of Quebec city (Société Dentaire de Québec). A short self-administered questionnaire was designed as a data collection instrument for dentist voluntarily responding to the invitation. Data collection was carried out from September 2016 to January 2017.

The database used for interrogation were Medline and Embase, from inception to 2022 with the following search terms: ultrasonic scalers, dental handpieces, occupational exposure or vibration exposure, dent* or dental personnel or oral health or dental specialists, dental hygienist, and dental laboratory technicians, musculoskeletal pain or disorders, hand-arm vibration syndrome, neuropathy, sensorineural disorder, and carpal tunnel syndrome. The eligibility criteria were the following: inclusion of relevant health outcomes, be published in French or English, original data (except one abstract in English), and human studies. The exclusion criteria were: lack of mentioning the prevalence of wrist/hand/fingers pain, ergonomic assessment of dentistry, preventive measures, lack of relation with the subject of study, letter to the editor, treatment, rehabilitation, perception of risks or awareness, experts statements on ergonomics in dentistry, and general topics on MSK and CTS. A concurrent search on vibration assessment of dental tools was performed.

3. Results

In total, 71 dentists (n = 71) consented to participate (women: 37, men: 30). Four incomplete questionnaires were rejected. The response rate was low (67/350). Years of practice varied from 0–10 years (n = 19), 11–20 y (n = 11), 21–30 y (n = 26), to more than 30 y (n = 11). Overall, 85% declared having exposure to handpieces for more than 11 h/week. Among the respondents, 14 dentists (13 women, 1 man) declare Raynaud's episodes (1 declare Raynaud's disease) during the last year, aged 35–44 y (n = 6), 45–54 y (n = 4), 55–64 y (n = 3), 65 and + (n = 1), cumulating years of practice as follows: 0–10 years (n = 2), 11–20 y (n = 3), 21–30 y (n = 8), and 31 and more (n = 1). Among them, two have white fingers without other symptoms. Four dentists report having consulted a physician for white fingers. Thirteen did not recall any restrictions in their professional tasks while six declared restrictions in their leisure activities. Among the respondents with episodes of white fingers, six reported having tingling, numbness, or tingling in the fingers, and seven have pain or stiffness in the finger joints. Two dentists report white fingers, tingling, numbness or tingling in the fingers, and pain or stiffness in the finger joints. Intolerance to cold, which manifests as pain, numbness, discomfort, and oedema of the fingers, afflicted 64% of respondents who report episodes of white fingers in the past year. Four of them consulted a doctor for this problem. Fifteen (22%) dentists described experiencing often or sometimes tingling and numbness in their fingers lasting more than 20 min/day. In total, 22% report (often or sometimes) tingling, numbness that wakes them up during the night. Among them, six dentists report having consulted a doctor for these problems. Twenty-four dentists (36%) reported pain or stiffness in the joints of the fingers and among them, four dentists treatment for this problem. Among these 24 dentists, 8 reported experiencing often or sometimes tingling and numbness in the fingers. In summary, 53% report one of the HAVS symptoms. Fourteen (14) of them (41.2%; 6 M, 8 F) report that their problems are work-related, of which nine dentists represent 21 years or more of experience, manipulating handpieces more than 10 h per week ($n = 11$).

3.1. Review of Literature

3.1.1. Vibration Assessment

Handpieces are mostly studied for their cutting and torque efficiency, noise [6–8], pinch force, or tool handle shape for non-vibrating curettes [9]. Information about vibration characteristics is sparse [10]. Five studies were retrieved. The assessment of vibration from dental tools dates back to 1979 [11]. Various methods of assessment of vibration have been described. Interpretation of results is controversial as low-frequency is known to be more important for adverse health effects according to frequency weighting of ISO 5349-1 standardized methods, despite generating high unweighted value [12]. Studies show that vibration spectra from handpieces and scalers is dominated by high-frequency component above 1000 Hz. Measurements during idling, normal work, or simulating drilling on plastic plate with air turbines or micromotor were performed with weighted vibration according to standard ISO-5349:1986, as well as the total acceleration of high-frequency unweighted. Weighted acceleration of 2–4 m/s^2 while grinding and from 3 to 500 m/s^2 above 1.2 kHz [13] or 0.01 to 0.2 m/s^2 during idling, values of 0.03 to 0.2 m/s^2 during drilling, while ultravibration during idling varied from 29 to 320 m/s^2 and during drilling from 120 to 640 m/s^2, old handpieces showing increased values [14]. During normal work, the weighted vibrations of air-turbine handpieces were 0.01–0.04 m/s^2 and for micromotor handpieces from 0.2 to 0.9 m/s^2. The range of ultravibration was 3–200 m/s^2 [15]. Using a laser vibrometer whilst idling, no single measurement for air turbines or micromotor exceeded 4 mm for the handpieces (in the range of 4–1607 m/s^2) [10]. A recent study measured the vibration amplitude of micromotor and air-turbine handpieces during idling although interpretation due to lack of methodology data is difficult [16]. In summary, vibration assessment of these tools is challenging due to their high frequency of oscillation and small associated displacement amplitude, technical problems relating to the adding mass of the accelerometer, replication of actual work during restorative procedures or periodontal work removing plaques, and frequency weighting [17].

3.1.2. Hand-Arm Vibration Syndrome and Related Disorders

A total of 891 articles were retrieved, of which 879 were kept, excluding duplicates. A total of 5 articles were rejected for plagiarism or error from the publisher. Sixteen references were added to make a total of 890 articles. Out of these, 535 were excluded and 355 were assessed according to pre-established criteria. There were a total of 36 literature reviews dealing with MSK disorders among dental professionals out of which international reviews and meta-analysis were retrieved [18–27]. One literature review on HAVS among dentists was found, as well as another review relating to neuropathy and high frequency vibration, and one on occupational hazards related to ultrasonic scalers [3,28–30]. Unfortunately, the review on HAVS retrieved six articles, of which one does not relate to dentistry.

The prevalence of musculoskeletal disorders is high among all dental professionals and has been studied worldwide, leading to serious impact on quality of life [22,26]. Females show a higher prevalence than males in some studies [18]. Use of dental tools are related to repetitive movements of the hand and fingers, pinch force, static and asymmetrical posture, precise hand movements, awkward postures of the wrist, high-frequency vibration, as well as other factors, such as poor visibility, lack of breaks between patients, and high job demand [31,32]. Despite the high prevalence of MSK among dental professionals, HAVS has been far less studied (Table A1). Studies with estimation of vibration exposure or risk assessment of vibration exposure according to ISO-5349-1 are sparse, as well as assessment of high-frequency vibration exposure among studies on hand/fingers pain.

For the outcome of Raynaud's phenomenon, 11 studies were included. White fingers have been described among dental hygienists, dentists, and dental technicians, with a prevalence varying ranged from 1.9% to 80%. Some studies present a positive association with daily use of high-frequency hand tools, lifetime exposure to vibration, total time dental filling, and root canal, while one found a negative association between Raynaud's phenomenon and a control group. Low cumulative exposure to vibration estimated at 1 m/s^2 during 8 h/day for 200 days/yr. or below the action value of 2.5 m/s^2 of the Vibration Directive of the European Union was described in two studies.

For the outcome neurosensory injury, 20 studies were included. Neurosensory injury has been described by questionnaire or objective neurophysiological tests. Significant impairments of vibrotactile sensibility, strength, and motor performance and more frequent sensorineural symptoms in the dominant hands of exposed dentists than controls were found. Injury to small nerve fibers and mechanoreceptors are suspected to affect tactile function. A high frequency of neurological symptoms among dentists with long-term experience especially in the dominant hand both in exposed and nonexposed fingers was found, suggesting other occupational risks than vibration. A positive association with years of experience and hours of using vibrating tools was described. The absence of a dental assistant and use of manually endodontic instruments were significantly related to a high prevalence of hand senso-neural disorders in root canal procedures compare to the use of low-speed dental handpieces, putting strain on the wrist due to pulling and pushing movements with the small handle of the instruments. Contrary these results, among dental hygienists and dental students, the vibration perception threshold in the 125 Hz (FAII receptors) indicate a fundamental difference between vibration exposure and the other biomechanical risk factors.

For the outcome of wrist/hand/fingers pain, 161 studies were included. The prevalence varied from 17% to 75% in the past 12 months among all dental professionals. In 18 studies, a positive association with vibrating tools or dental work was found, such as number of hours using vibrating tool, procedures such as scaling, inability to select the size of dental tools, or time spent forcefully gripping a tool with each patient or time. High job demand and stress was associated with symptoms in wrist/hand in three studies.

For the outcome of decreased grip strength, six studies were included. For the outcome of osteoarthritis (OA) of the fingers, six studies were included. OA in the distal interphalangeal joints among dentists without statistical difference between the drill and mirror hand contrary to another study showing more severe OA in the right-hand thumb and the index and middle fingers and among dentists with a low task variation was found. Arthrosis of the upper extremities without significant difference among dentists and dental assistants was reported. The results from the questionnaire showed thumb disability among female dentists aged 50 years and over. Upon physical exam, in one study, Heberden's arthrosis among dentists, dental hygienists, and assistants ($n = 15$) was compared to referents ($n = 6$). For the outcome of carpal tunnel syndrome, 41 studies were included. The prevalence varies from 3.8% to 86% by questionnaire and a lower prevalence was shown with nerve conduction studies. The prevalence in the dominant hand is not always reported. The prevalence seems higher among dental hygienists. The assessment of high-frequency vibration as an occupational risk is seldom reported. Significant risk factors of CTS were found: (1) vibration exposure greater than two hours per day, a wrist diameter ratio of greater than 0.7, and female sex; (2) an association of vibration exposure and two different pathologies associated with paresthesia, such as CTS and weak pinch grip associated with raised vibration perception threshold; and (3) specific dental procedures due to repetitive movements performed during scaling or to vibration of the handpieces and scalers during shaping of the root canal, polishing, or removing heavy calculus.

4. Discussion and Conclusions

Dental professionals are exposed to a vast array of occupational risks while exposed to vibrating handpieces. Neurosensory injury is more frequently described than Raynaud's phenomenon, documented with neurophysiologic tests among studies. Only one study reported results of the cold provocation test. The estimates of prevalence of Raynaud's, according to the daily equivalent vibration expressed as A(8), are missing. Diminished grip, osteoarthritis, carpal tunnel, and hand/finger pain have been described, but there is a lack of information on dental procedures among dental professionals to fully understand the association between all occupational risks and diseases. Vibration assessment of high-frequency handpieces is mainly reported from old studies, relying on metrics giving more importance to low-frequency vibration. Measurement of daily vibration exposure constructed with the frequency weighting Wp (Ap(8) and Ep,d), giving more importance to intermediate and high-frequency vibration, would provide better assessment of occurrence of Raynaud's phenomenon than the metric derived from the conventional ISO frequency weighting Wh (Ah(8)) [33]. Our questionnaire survey shows that HAVS may be present among dentists in accordance with our literature review. A more in-depth evaluation would rule out other probable causes of paresthesia or Raynaud's phenomenon. The consultation of four respondents' dentists with a physician deserves attention. Future studies need to be performed with a protocol well designed for HAVS research among dental workers. High-frequency vibration from handpieces and scalers is absorbed by superficial tissues and tissue structures affecting mechanoreceptors sensitive to mechanical skin displacement (Pacinian responding to 5–800 Hz, less likely Meisner corpuscules 1–300 Hz, Merkel cells 100 Hz), as well as small nerve fibers [34–37]. Lundstrom suspected that if the signals from these receptors are disturbed, a reduction in tactile sensitivity would be detectable at lower frequencies, as shown in [11]. Surprisingly, these high frequencies fall out of the resonant frequency of the human hand (100–300 Hz), which generate an increased biodynamic response in the exposed tissue [38]. Hand intensive work, local stress, and high pinch force can also constitute contributory factors for the occurrence of vascular and nerve damage and osteoarthritis [39]. Repetitive movements and awkward and static posture could cause nerve compression at the wrist and increase the risk for CTS. Our results suggest that the usage of high- and low-speed type dental handpieces represent an occupational risk among dental professionals deserving further study.

Author Contributions: Conceptualization, A.T., D.H. and M.T.; methodology, A.T., D.H., and M.T.; validation, formal analysis, writing—original draft preparation, writing—review and editing: A.T. All authors have read and agreed to the published version of the manuscript.

Funding: This research received no external funding.

Institutional Review Board Statement: The survey was approved by the scientific committee of the Société Dentaire de Québec and approved by the Institut national de santé publique.

Informed Consent Statement: Informed consent was obtained from all subjects involved in the study.

Data Availability Statement: The survey's data are available on request from the corresponding author, the data are not publicly available due to preliminary step of the study.

Conflicts of Interest: The authors declare no conflict of interest.

Appendix A

Table A1. HAVS and Dentists.

Author	Population Study	Findings	Tests/Questionnaire
Lundström and Lindmark [11]	Dentists ($n = 10$), controls ($n = 10$)	Neuropathy	Vibrogram: reduction of vibration perception
Hjortsberg [13]	Dental technicians ($n = 10$), controls ($n = 10$)	HAVS: SV Stockholm: $n = 4$ SN Stockholm: $n = 8$	Questionnaire, sensory nerve conduction, abnormal vibrogram, warming & cold thresholds
Ekenvall [4]	Dentists: long term ($n = 26$), short term ($n = 18$)	Neuropathy, other causes suspected	Vibrogram, temperature & pain threshold, daily exposure to dental tools difficult to estimate
Milerad and Ekenvall [5]	Dentists: ($n = 99$), controls ($n = 100$)	Vascular symptoms RR:1.2 (CI 95%: 0.7–49.5) & neurological symptoms: RR:4.2 (CI 95%: 2.3–7.7)	Questionnaire, women had higher Raynaud's phenomenon, unilateral Raynaud's in dentist in the dominant hand of dentists
Yoshida [40]	Dental technicians ($n = 164$)	HAVS: 5.5% vascular symptoms & 18.5%: numbness	Symptoms related to the working posture and used of dental vibrating tools
Stockhill [41]	Dentists ($n = 1016$ respondents)	Neurological symptoms: 15% numbness,17% tingling, 25% pain	Questionnaire
Stentz [42]	Dental hygiene ($n = 260$)	Neurological symptoms: 61% (pain, tingling, numbness)	Questionnaire
Akesson [43]	Females dentists & dental hygienists; dental assistants & controls ($n = 30$ in each group)	HAVS: Neurological symptoms ($n = 18/90$) and vascular symptoms ($n = 16/90$), decreased hand grip, no increase of vascular symptoms in the groups exposed to vibration. Abnormal neurological tests	Neurological tests (vibrogram, sensibility index, two point discrimination, tactile identification test, grip strength)
Nakladalova [44]	Dental technicians ($n = 120$)	HAVS: numbness: 52.5%, white fingers: 0.03%, abnormal plethysmographic curves & abnormal EMG: $n = 13/54$ subjects tested	Questionnaire. Cold water test, plethysmographic investigation, EMG, X-ray, neurological and orthopaedic exam
Keruoso [45]	Dentists ($n = 147$), orthodentists ($n = 81$), controls ($n = 99$)	HAVS neurological symptoms (10.6%) & vascular symptoms (1.9%)	Questionnaire
Alnaser [46]	Dentists ($n = 89$)	Wrist pain, 1 case of HAVS/89 respondents	Questionnaire
Bylund [47]	Females dental hygienists ($n = 21$), dentists ($n = 26$), technicians ($n = 31$)	HAVS vascular symptoms: respectively (38%), (60%), (60%) & neurological symptoms (94%), (79%), (92%), decreased hand grip	Questionnaire
Morse [48]	Dental hygiene students ($n = 82$)	HAVS numbness (13%) & white fingers or painful fingers in cold (13%)	Numbness & tingling increase with each hour per week vibrating tools OR:1.10 (CI 95%: 1.01-1.19)
Cherniak [49]	Dental hygienists ($n = 94$) & students ($n = 66$)	HAVS: sensorineural (Stockholm SK): 45% hygienists, dental hygiene students: 9%, vascular symptoms:12% among both groups	Questionnaire, vibrogram, nerve conduction study, pinch & grip force, neither manual or vibratory accounted for VPT thresholds
Gijbels [36]	Dentists ($n = 20$ randomly selected /500)	HAVS Neurological symptoms (6%) decreasing discrimination with years of practice	Questionnaire, two points discrimination, thermal sensory test, light touch
Rytkönen [15]	Dentists ($n = 295$ females)	Fingers symptoms related with total time dental filling &root canal OR:.1.9 (CI 95%: 1.03–3.6)	Questionnaire & pinch strength, symptoms not specified, grip strength inversely related to finger symptoms

Table A1. *Cont.*

Author	Population Study	Findings	Tests/Questionnaire
Morse [50]	Dental hygienists (n=27) & dental hygiene students (n = 39)	Physical exam: no HAVS found	-
Shabazian [37]	Dentists (n = 50) & controls (n = 20)	Neuropathy: diminished tactile sensibility in dentist with >25 years of practice	Light touch, two points discrimination, thermal sensation of both hand: reduction of tactile sensibility
Bjorkman [51]	Dental technicians (n = 10) & controls (n = 10)	Cortical reorganization in dental technician's group with neuropathy than in controls	Functional magnetic resonance
Warren [52]	Dental hygienists (n = 94) & dental Hygiene students (n = 66)	Neuropathy and Carpal tunnel syndrome	Vibrogram & nerve conduction studies, role of vibrating tools
Ancuta [53]	Dentists (n = 30)	Neurological symptoms: paresthesias and thenar amyotrophy (n = 3/30)	Cornell musculoskeletal discomfort Questionnaire & exam
Zoidaki [54]	Dentists (n = 80)	Neurological symptoms, women reported more sensorineural disorders OR: 2.6 (CI 95%: 1.06–6.7)	Nordic Musculoskeletal questionnaire, Sensorineural symptoms in root canal (manual vs. rotor) - OR: 3.4 (CI 95%: 1.08–10.9)
Jaque and Burke [55]	Dentists (n = 6/10)	HAVS vascular 80%, numbness 80%, tingling 10% & pain 60%	Questionnaire

References

1. Little, D. *Handpieces and Burs: The Cutting Edge*; PennWell: Tulsa, OK, USA, 2014.
2. Chen, Y.L.; Chang, H.H.; Chiang, Y.C.; Lin, C.P. Application and development of ultrasonics in dentistry. *J. Formos. Med. Assoc.* **2013**, *112*, 659–665. [CrossRef]
3. Arabaci, T.; Çiçek, Y.; Çanakçi, C. Sonic and ultrasonic scalers in periodontal treatment: A review. *Int. J. Dent. Hyg.* **2007**, *5*, 2–12. [CrossRef] [PubMed]
4. Ekenvall, E.; Nilsson, B.Y.; Falconer, C. Sensory perception in the hands of dentists. *Scand. J. Work Environ. Health* **1990**, *16*, 334–339. [CrossRef]
5. Milerad, E.; Ekenvall, L. Symptoms of the neck and upper extremities in dentists. *Scand. J. Work Environ. Health* **1990**, *16*, 129–134. [CrossRef]
6. Choi, C.; Driscoll, C.F.; Romberg, E. Comparison of cutting efficiencies between electric and air-turbine dental handpieces. *J. Prosthet. Dent.* **2010**, *103*, 101–107. [CrossRef] [PubMed]
7. Sorainen, E.; Rytkönen, E. Noise level and ultrasound spectra during burring. *Clin. Oral Investig.* **2002**, *6*, 133–136. [CrossRef]
8. Henneberry, K.; Hilland, S.; Haslam, S.K. Are dental hygienists at risk for noise-induced hearing loss? A literature review. *Can. J. Dent. Hyg.* **2021**, *55*, 110–119.
9. Dong, H.; Loomer, P.; Barr, A.; LaRoche, C.; Young, E.; Rempel, D. The effect of tool handle shape on hand muscle load and pinch force in a simulated dental scaling task. *Appl. Ergon.* **2007**, *38*, 525–531. [CrossRef] [PubMed]
10. Poole, R.L.; Lea, S.C.; Dyson, J.E.; Shortall, A.C.C.; Walmsley, A.D. Vibration characteristics of dental high-speed turbines and speed-increasing handpieces. *J. Dent.* **2008**, *36*, 488–493. [CrossRef] [PubMed]
11. Lundström, R.; Lindmark, A. Effects of Local Vibration on Tactile Perception in the Hands of Dentists. *J. Low Freq. Noise Vib. Act. Control* **1982**, *1*, 1–11. [CrossRef]
12. Mansfield, N.J. The European vibration directive–how will it affect the dental profession? *Br. Dent. J.* **2005**, *199*, 575–608. [CrossRef]
13. Hjortsberg, U.; Rosen, I.; Orbaek, P.; Lundborg, G.; Balogh, I. Finger receptor dysfunction in dental technicians exposed to high-frequency vibration. *Scand. J. Work. Environ. Health* **1989**, *15*, 339–344. [CrossRef] [PubMed]
14. Rytkönen, E.; Sorainen, E. Vibration of Dental Handpieces. *Am. Ind. Hyg. Assoc.* **2001**, *62*, 477–481. [CrossRef]
15. Rytkönen, E.; Sorainen, E.; Leino-Arjas, P.; Solovieva, S. Hand-arm vibration exposure of dentists. *Int. Arch. Occup. Environ. Health* **2006**, *79*, 521–527. [CrossRef] [PubMed]
16. Banga, H.K.; Goel, P.; Kumar, R.; Kumar, V.; Kalra, P.; Singh, S.; Singh, S.; Prakash, C.; Pruncu, C. Vibration Exposure and Transmissibility on Dentist's Anatomy: A Study of Micro Motors and Air-Turbines. *Int. J. Environ. Res. Public Health* **2021**, *18*, 4084. [CrossRef]
17. Lea, S.C.; Landini, G.; Walmsley, A.D. Vibration characteristics of ultrasonic scalers assessed with scanning laser vibrometry. *J. Dent.* **2002**, *30*, 147–151. [CrossRef] [PubMed]
18. Chenna, D.; Pentapati, K.C.; Kumar, M.; Madi, M.; Siddiq, H. Prevalence of musculoskeletal disorders among dental healthcare providers: A systematic review and meta-analysis. *F1000Research* **2022**, *11*, 1062. [CrossRef]
19. Hayes, M.J.; Smith, D.R.; Cockrell, D. An international review of musculoskeletal disorders in the dental hygiene profession. *Int. Dent. J.* **2010**, *60*, 343–352. [CrossRef]
20. Hayes, M.; Cockrell, D.; Smith, D.R. A systematic review of musculoskeletal disorders among dental professionals. *Int. J. Dent. Hyg.* **2009**, *7*, 159–165. [CrossRef]
21. Lietz, J.; Ulusoy, N.; Nienhaus, A. Prevention of musculoskeletal diseases and pain among dental professionals through ergonomic interventions: A systematic literature review. *Int. J. Environ. Res. Public Health* **2020**, *17*, 3482. [CrossRef]
22. Lietz, J.; Kozak, A.; Nienhaus, A. Prevalence and occupational risk factors of musculoskeletal diseases and pain among dental professionals in Western countries: A systematic literature review and meta-analysis. *PLoS ONE* **2018**, *13*, e0208628. [CrossRef]
23. Soo, S.Y.; Ang, W.S.; Chong, C.H.; Tew, I.M.; Yahya, N.A. Occupational ergonomics and related musculoskeletal disorders among dentists: A systematic review. *Work* **2022**, *74*, 469–476. [CrossRef]
24. Sultana, N.; Mian, M.A.H.; Rubby, M.G.; Banik, P.C. Musculoskeletal Disorders in Dentists: A Systematic Review, Update. *Dent. Coll. J.* **2018**, *7*, 38–42. [CrossRef]
25. AlOtaibi, F.; Nayfeh, F.M.M.; Alhussein, J.I.; Alturki, N.A.; Alfawzan, A.A. Evidence based analysis on neck and low back pain among dental practitioners- a systematic review. *J. Pharm. Bioallied Sci.* **2022**, *14*, S897–S902. [CrossRef]
26. ZakerJafari, H.R.; YektaKooshali, M.H. Work-Related Musculoskeletal Disorders in Iranian Dentists: A Systematic Review and Meta-analysis. *Saf. Health Work* **2018**, *9*, 1–9. [CrossRef]
27. ShamsHosseini, N.; Vahdati, T.; Mohammadzadeh, Z.; Yeganeh, A.; Davoodi, S. Prevalence of Musculoskeletal Disorders among Dentists in Iran: A Systematic Review. *Mater. Socio-Med.* **2017**, *29*, 257–262. [CrossRef] [PubMed]
28. Chowdhry, R.; Sethi, V. Hand arm vibration syndrome in dentistry: A review. *Curr. Med. Res. Pract.* **2017**, *7*, 235–239. [CrossRef]
29. Szymanska, J. Dentist's hand symptoms and high-frequency vibration. *Ann. Agric. Environ. Med.* **2001**, *8*, 7–10. [PubMed]
30. Trenter, S.C.; Walmsley, A.D. Ultrasonic dental scaler: Associated hazards: Ultrasonic dental scaler. *J. Clin. Periodontol.* **2003**, *30*, 95–101. [CrossRef]

31. De Sio, S.; Traversini, V.; Rinaldo, F.; Colasanti, V.; Buomprisco, G.; Perri, R.; Mormone, F.; La Torre, G.; Guerra, F. Ergonomic risk and preventive measures of musculoskeletal disorders in the dentistry environment: An umbrella review. *PeerJ* **2018**, *6*, e4154. [CrossRef]
32. Ohlendorf, D.; Naser, A.; Haas, Y.; Haenel, J.; Fraeulin, L.; Holzgreve, F.; Erbe, C.; Betz, W.; Wanke, E.M.; Brueggmann, D.; et al. Prevalence of musculoskeletal disorders among dentists and dental students in Germany. *Int. J. Environ. Res. Public Health* **2020**, *17*, 8740. [CrossRef] [PubMed]
33. Bovenzi, M. Epidemiological evidence for new frequency weightings of hand-transmitted vibration. *Ind. Health* **2012**, *50*, 377–387. [CrossRef]
34. Johnson, K.O.; Yoshioka, T.; Vega–Bermudez, F. Tactile Functions of Mechanoreceptive Afferents Innervating the Hand. *J. Clin. Neurophysiol.* **2000**, *17*, 539–558. [CrossRef]
35. Deflorio, D.; Di Luca, M.; Wing, A.M. Skin and Mechanoreceptor Contribution to Tactile Input for Perception: A Review of Simulation Models. *Front. Hum. Neurosci.* **2022**, *16*, 862344. [CrossRef]
36. Gijbels, F.; Jacobs, R.; Princen, K.; Nackaerts, O.; Debruyne, F. Potential occupational health problems for dentists in Flanders, Belgium. *Clin. Oral Investig.* **2006**, *10*, 8–16. [CrossRef]
37. Shahbazian, M.; Bertrand, P.; Abarca, M.; Jacobs, R. Occupational changes in manual tactile sensibility of the dentist. *J. Oral Rehabil.* **2009**, *36*, 880–886. [CrossRef]
38. Krajnak, K. Health effects associated with occupational exposure to hand-arm or whole body vibration. *J. Toxicol. Environ. Health Part B Crit. Rev.* **2018**, *21*, 320–334. [CrossRef]
39. Nilsson, T.; Wahlstrom, J.; Burstrom, L. Hand-arm vibration and the risk of vascular and neurological diseases—A systematic review and meta-analysis. *PLoS ONE* **2017**, *12*, e0180795. [CrossRef] [PubMed]
40. Yoshida, H.; Nagata, C.; Mirbod, S.M.; Iwata, H.; Inaba, R. Analysis of subjective symptoms of upper extremities in dental technicians. *Jpn. J. Ind. Health* **1991**, *33*, 17–22. [CrossRef]
41. Stockstill, J.W.; Harn, S.D.; Strickland, D.; Hruska, R. Prevalence of upper extremity neuropathy in a clinical dentist population. *J. Am. Dent. Assoc.* **1993**, *124*, 67–72. [CrossRef] [PubMed]
42. Stentz, T.L.; Riley, M.W.; Harn, S.D.; Sposato, R.C.; Stockstill, J.W.; Harn, J.A. Upper extremity altered sensations in dental hygienists. *Int. J. Ind. Ergon.* **1994**, *13*, 107–112. [CrossRef]
43. Akesson, I.; Lundborg, G.; Horstmann, V.; Skerfving, S. Neuropathy in female dental personnel exposed to high frequency vibrations. *Occup. Environ. Med.* **1995**, *52*, 116–123. [CrossRef]
44. Nakladalova, M.; Fialova, J.; Korycanova, H.; Nakladal, Z. State of health in dental technicians with regard to vibration exposure and overload of upper extremities. *Cent. Eur. J. Public Health* **1995**, *3*, 129–131. [PubMed]
45. Kerosuo, E.; Kerosuo, H.; Kanerva, L. Self-reported health complaints among general dental practitioners, orthodontists, and office employees. *Acta Odontol. Scand.* **2000**, *58*, 207–212. [PubMed]
46. Alnaser, M.Z.; Almaqsied, A.M.; Alshatti, S.A. Risk factors for work-related musculoskeletal disorders of dentists in Kuwait and the impact on health and economic status. *Work* **2021**, *68*, 213–221. [CrossRef]
47. Bylund, S.H.; Burstrom, L.; Knutsson, A. A descriptive study of women injured by hand-arm vibration. *Ann. Occup. Hyg.* **2002**, *46*, 299–307. [CrossRef]
48. Morse, T.F.; Michalak-Turcotte, C.; Atwood-Sanders, M.; Warren, N.; Peterson, D.R.; Bruneau, H.; Cherniack, M. A pilot study of hand and arm musculoskeletal disorders in dental hygiene students. *J. Dent. Hyg.* **2003**, *77*, 173–179.
49. Cherniack, M.; Brammer, A.J.; Nilsson, T.; Lundstrom, R.; Meyer, J.D.; Morse, T.; Neely, G.; Peterson, D.; Toppila, E.; Warren, N. Nerve conduction and sensorineural function in dental hygienists using high frequency ultrasound handpieces. *Am. J. Ind. Med.* **2006**, *49*, 313–326. [CrossRef]
50. Morse, T.; Bruneau, H.; Michalak-Turcotte, C.; Sanders, M.; Warren, N.; Dussetschleger, J.; Diva, U.; Croteau, M.; Cherniack, M. Musculoskeletal disorders of the neck and shoulder in dental hygienists and dental hygiene students. *J. Dent. Hyg.* **2007**, *81*, 10.
51. Bjorkman, A.; Weibull, A.; Svensson, J.; Balogh, I.; Rosen, B. Cortical changes in dental technicians exposed to vibrating tools. *NeuroReport* **2010**, *21*, 722–726. [CrossRef]
52. Warren, N. Causes of musculoskeletal disorders in dental hygienists and dental hygiene students: A study of combined biomechanical and psychosocial risk factors. *Work* **2010**, *35*, 441–454. [CrossRef] [PubMed]
53. Ancuta, C.; Iordache, C.; Fatu, A.M.; Alunculesei, C.; Forna, N. Ergonomics and prevention of musculoskeletal-work related pathology in dentistry: A pilot study. *Rom. J. Oral Rehabil.* **2016**, *8*, 73–79.
54. Zoidaki, A.; Riza, E.; Kastania, A.; Papadimitriou, E.; Linos, A. Musculoskeletal disorders among dentists in the Greater Athens area, Greece: Risk factors and correlations. *J. Public Health (Germany)* **2013**, *21*, 163–173. [CrossRef]
55. Burke, F.J.; Jaques, S.A. Vibration white finger. *Br. Dent. J.* **1993**, *174*, 194. [CrossRef] [PubMed]

Proceeding Paper

Hand-Arm Vibration Exposure Trends among the Workforce in Sweden [†]

Hans Pettersson [1,*], Mattias Sjöström [2,3] , Max Wikström [3] and Jenny Selander [3]

[1] Department of Public Health and Clinical Medicine, Medical Faculty, Umea University, 907 37 Umea, Sweden
[2] Center for Occupational Environmental Medicine (CAMM), Region Stockholm, 113 65 Stockholm, Sweden; mattias.sjostrom@ki.se
[3] Institute of Environmental Medicine (IMM), Karolinska Institutet, 171 77 Stockholm, Sweden; max.wikstrom@ki.se (M.W.); jenny.selander@ki.se (J.S.)
[*] Correspondence: hans.pettersson@umu.se; Tel.: +46-907856927
[†] Presented at the 15th International Conference on Hand-Arm Vibration, Nancy, France, 6–9 June 2023.

Abstract: The aim of this research was to study hand-arm vibration (HAV) exposure trends in the workforce in Sweden by using a Job-Exposure matrix (JEM). All individuals employed during 1980 and 2010 with an occupational code were included. The daily eight-hour equivalent HAV exposure values were divided into three exposure categories. During the study period, the proportion of workers exposed above the action value had decreased, as well as the proportion of workers exposed to any HAV. In 2010, 4% of the workforce in Sweden were exposed to HAV above the action value.

Keywords: hand-arm vibration; job-exposure matrix trend; workforce

Citation: Pettersson, H.; Sjöström, M.; Wikström, M.; Selander, J. Hand-Arm Vibration Exposure Trends among the Workforce in Sweden. *Proceedings* **2023**, *86*, 2. https://doi.org/10.3390/proceedings2023086002

Academic Editors: Christophe Noël and Jacques Chatillon

Published: 6 April 2023

1. Introduction

Hand-arm vibration (HAV) is common among construction, industry, forestry, and manufacturing workers. HAV could cause vascular, neurological, and musculoskeletal injuries over time [1]. It is therefore important to decrease the exposure levels. To study changes in HAV exposure over time, a Job-Exposure matrix was constructed for the Swedish workforce. The JEM consisted of the eight-hour equivalent HAV exposure connected to each occupational code. More participants could be included by using a JEM, and it could be used to study trends regarding the workforce. The aim of this paper was to study trends of HAV exposure from 1980 to 2010.

2. Materials and Methods

A worker was included if they were above 18 years old, employed for one year in the Swedish workforce during the study period of 1980–2010, and had a occupation with an occupational code. Information on age was gathered from the Register on the Entire Population (Registret över totalbefolkningen) at Statistics Sweden. Information on employment and job title among the Swedish workforce was gathered from FOB 1980 and 1990 and SSYK-96. Every occupation in Sweden is coded according to the occupational classifications of the National Labour Market Board (Arbetsmarknadsstyrelsens yrkesklassificering). The occupational classification code is based on the International Standard Classification of Occupations, ISCO-88-code system, and has been described elsewhere [2]. The FOB80 classification was used for exposures between 1980 and 1990, and SSYK96 was used for 2001–2013 with four-digit codes. The eight-hour equivalent HAV exposure level, A(8), was calculated for each occupational code and for each job classification system. The A(8) was calculated according to the present international standard ISO 5349-1. The HAV levels were based on earlier measurements from scientific articles, measurement reports, and vibration databases (n = 90). The exposure categories for the A(8) value were low (range: above 0 to <=1 m/s^2), moderate (range: above 1 to <2.5 m/s^2), and high (range: =>2.5). HAV exposure in the high group was set according to exposure above the action value from the EU directive on vibration [3].

3. Results

During the period 1980–2010, the proportion of workers exposed to high HAV levels decreased (Table 1). From 1980 to 2010, the proportion of workers in the low and moderate HAV exposed groups also decreased. In 2010, the proportion of workers with any HAV exposure decreased from 29% to 19%.

Table 1. Proportion of workers with daily equivalent hand-arm vibration exposure (low (range: above 0 to <=1 m/s^2), moderate (range: above 1 to <2.5 m/s^2), and high (range: =>2.5)) from 1980 to 2010.

HAV	1980	1990	2001	2010
Low	12	11	10	10
Moderate	7	6	5	5
High	10	8	4	4
All	29	25	19	19

4. Discussion

The proportion of workers exposed to high HAV levels in Sweden declined during the study period. During this time, an EU directive was implemented to reduce workers' exposure to HAV [3]. Some machines have better designs to reduce the vibration levels. Additionally, there are less workers in occupations in which they use hand-held vibrating tools, since manufacturing declined in Sweden during the study period.

Author Contributions: Conceptualization, H.P., M.S., and J.S.; methodology, H.P., J.S., M.S., and M.W.; software, J.S.; validation, H.P.; formal analysis, J.S., M.W., and H.P.; investigation, H.P., M.S., and J.S.; resources, J.S.; data curation, J.S. and H.P.; writing—original draft preparation, H.P. and J.S.; writing—review and editing, H.P., J.S., M.S., and M.W.; visualization, H.P.; supervision, J.S.; project administration, J.S.; funding acquisition, J.S. All authors have read and agreed to the published version of the manuscript.

Funding: This research was funded by The Swedish Research Council for Health, Working Life and Welfare (FORTE), grant number 2016-07185.

Institutional Review Board Statement: The study was conducted in accordance with the Declaration of Helsinki and approved by The Ethical Review Board in Stockholm (protocol code 2018/1007-31/5).

Informed Consent Statement: Subject consent was waived due to use of Swedish public register data which allow universities in Sweden to conduct their own evaluations regarding consent and the use of data without consent if it will benefit the citizens.

Data Availability Statement: Information and data from the Job-Exposure matrix used will be available from February 2023 on the Karolinska Institutet homepage.

Acknowledgments: We appreciate the support in the construction of the JEM for HAV and the help of Bodil Björ at the Occupational and Environmental Medicine at the Region Västerbotten in Umeå, Sweden.

Conflicts of Interest: The authors declare no conflict of interest. The funders had no role in the design of the study; in the collection, analyses, or interpretation of data; in the writing of the manuscript; or in the decision to publish the results.

References

1. Nilsson, T.; Wahlstrom, J.; Burstrom, L. Hand-arm vibration and the risk of vascular and neurological diseases-A systematic review and meta-analysis. *PLoS ONE* **2017**, *12*, e0180795. [CrossRef] [PubMed]
2. Selander, J.; Albin, M.; Rosenhall, U.; Rylander, L.; Lewné, M.; Gustavsson, P. Maternal Occupational Exposure to Noise during Pregnancy and Hearing Dysfunction in Children: A Nationwide Prospective Cohort Study in Sweden. *Environ. Health Perspect.* **2015**, *124*, 855–860. [CrossRef] [PubMed]
3. EU. Directive 2002/44/EC of the European Parliament and the Council of the European Union on the minimum health and safety requirements regarding the exposure of workers to the risks arising from physical agents (vibration). *Off. J. Eur. Communities* **2002**, *117*, 6–7.

proceedings

Proceeding Paper

A Delphi Study to Address a Number of Issues Relating to the Practical Management of Hand-Arm Vibration Syndrome and Carpal Tunnel Syndrome in the Workplace †

Roger Cooke *, Dan Ashdown, Harriet Fox, Cornelius Grobler, Rupert Hall-Smith, Dominic Haseldine, Elschen Kotze and Ian Lawson

HAVS Delphi Study of the (UK) Society of Occupational Medicine Special Interest Group in Hand Arm Vibration Syndrome, London NW1 4LB, UK
* Correspondence: drrogercooke@icloud.com
† Presented at the 15th International Conference on Hand-Arm Vibration, Nancy, France, 6–9 June 2023.

Abstract: A Delphi study has been undertaken to address eight specific areas relating to the management of hand-arm vibration syndrome and carpal tunnel syndrome, with the aim of providing consensus guidelines.

Keywords: Raynaud's; carpal tunnel; sensory testing; standardised testing; Dupuytren's; health surveillance

Citation: Cooke, R.; Ashdown, D.; Fox, H.; Grobler, C.; Hall-Smith, R.; Haseldine, D.; Kotze, E.; Lawson, I. A Delphi Study to Address a Number of Issues Relating to the Practical Management of Hand-Arm Vibration Syndrome and Carpal Tunnel Syndrome in the Workplace. *Proceedings* **2023**, *86*, 43. https://doi.org/10.3390/proceedings2023086043

Academic Editors: Christophe Noël and Jacques Chatillon

Published: 17 May 2023

1. Introduction

It is over 100 years since the relationship between vibration exposure and symptoms affecting the hands was first recognised. In that time, there have been developments in the approach to staging the severity of these symptoms, which are largely reliant on staging systems, such as the Taylor–Pelmear, and latterly, the modified Stockholm Scale. In the UK, regulations were introduced in 2005, with associated guidance on the assessment of risks, the process of health surveillance, and the management of affected employees. However, a number of issues relating to the management of such employees remain poorly defined.

The Society of Occupational Medicine (SOM) Special Interest Group (SIG) was established in 2017 to facilitate discussion relating to any aspect of vibration-related diseases among members with a particular interest and/or expertise. Since then, a number of publications have been produced, addressing a range of associated topics, but it became increasingly apparent that there was a range of sometimes markedly divergent opinions with regard to several issues. It is believed that this divergence of opinion is not only representative of UK practitioners.

With the aim of providing a consensus opinion on a number of these issues, it was agreed that a Delphi Group should be established [1].

2. Aims

The aim of the Delphi study is to review a number of specific issues that are related to hand arm vibration syndrome (HAVS), about which there is no definitive evidence, but for which a consensus view would be likely to assist those undertaking HAVS surveillance and assessments.

3. Method

In total, 15 members of the SOM SIG agreed to participate, all being occupational physicians with experience of hand-arm vibration syndrome. It was agreed that eight specific topic areas would be subject to the Delphi process, undertaken by email, with one member of the group acting as a moderator for each topic. Following the completion of the exercise, the eight topics have been combined to present this summary consensus paper. The broad topics considered are in Table 1, with specific statements subsequently designed by the topic moderator in a format that is consistent with a Delphi exercise, in order to allow for agreement or disagreement and a presentation of supportive evidence from each participant. The moderator formulated the statement(s) for each round, such that the responses are "agree/disagree/undecided".

Table 1. Subjects streams and initial questions for Delphi study.

Set	Topic	Issues to Be Addressed
1	Primary Raynaud's phenomenon (RP)	What criteria should be used to differentiate primary RP from vascular HAVS? What advice should be offered to those with primary RP wishing to work with exposure to hand transmitted vibration (HTV)? What criteria should lead to a referral for a further investigation of RP?
2	Frequency of health surveillance	How frequent should increased surveillance be performed for those with stage 2 HAVS and how long should the increased frequency of surveillance continue?
3	Criteria for vascular staging	With vascular HAVS, should the extent of the blanching always override the frequency of the blanching when staging? If not, how do you balance the frequency and extent when grading?
4	Use of monofilaments for sensory testing	What cut-off of WEST/SW monofilaments should be used to assess normal sensory perception when assessing whether reduced sensory perception is present in those exposed to hand transmitted vibration? What other factors should be considered when interpreting the results of monofilament testing?
5	Carpal tunnel Syndrome (CTS)	Should cases of suspected CTS from history and examination be referred for nerve conduction studies before confirming a diagnosis? Should cases of suspected CTS be restricted from using hand vibrating tools until an investigation and treatment is completed? Should cases of a recurrence of CTS be permanently restricted from using vibrating tools?
6	Peripheral neuropathy and sensorineural HAVS	What advice should be offered to those with peripheral neuropathy/neurological symptoms similar to HAVS that are wishing to work with exposure to HTV? Is there an overlap of HAVS SN symptoms with diabetic neuropathy (DN) symptoms when performing HAV surveillance? What should the frequency of surveillance be? How to mitigate the legal risks for an employer with a missed diagnosis of HAVS masked by DN symptoms?
7	Dupuytren's disease	Should cases of Dupuytren's contracture be restricted from using vibrating tools? If yes, to what severity?
8	The use of quantitative tests for routine health surveillance	When should cases of HAVS be referred for a tier 5 assessment? Should reduced sensory perception in sensory HAVS be assessed by using more than one QST? If so, at what stage should ST be considered?

4. Results

There was a good consensus regarding issues relating to differentiation of primary Raynaud's phenomenon (PRP) from hand arm vibration syndrome (HAVS), and the management of cases of PRP in the workplace. It was agreed through separate statements that PRP generally shows an age of onset below 30 years, usually presents with symmetrical blanching, and that a positive family history and involvement of feet and/or other peripheries is indicative of PRP rather than HAVS. Vascular HAVS generally results from significant vibration exposure, so alternative diagnoses including PRP should be considered in those with short duration lifetime exposure (less than 5 years). Conversely there was a consensus agreement that asymmetrical blanching primarily involving the trigger fingers and leading hand would be more suggestive of HAVS than PRP. Symmetrical blanching affecting all fingers of both hands (with or without other extremities) warrants more in depth enquiry to exclude other conditions (e.g., autoimmune disease, blood or vascular disorders, medication etc.) when it presents in vibration exposed individuals over the age of 30, with no family history of PRP. HTV exposed individuals with a history of blanching and possible Carpal Tunnel Syndrome should be referred for investigation and treatment of CTS prior to diagnosing RP or vascular HAVS.

There was agreement regarding the management of PRP in the workplace, including that in those with known PRP, exposure to hand-transmitted vibration should be kept as low as practicable below the EAV of 2.5 m/s^2 or 100 points on the HSE scale, and that these employees should be subject to enhanced surveillance to include annual review of photographic evidence to help monitor progression of symptoms.

HTV exposed individuals who are diagnosed with PRP at health surveillance, should be advised that they can continue with limited exposure (below the Exposure Action Value) with careful monitoring. Those with blanching and a history of health issues known to be associated with RP (e.g., scleroderma, connective tissue disorders, rheumatoid arthritis, hypothyroidism) should be referred.

In respect of HAVS there was 100% agreement that with vascular HAVS the extent of blanching should over-ride frequency of attacks, and that, while photographic evidence should be used to confirm the diagnosis and extent of blanching and vascular staging, the absence of photographic evidence should not be used to discount or overturn a presumptive diagnosis of HAVS where there is a history of sufficient exposure and anamnesis of cold induced distal circumferential finger blanching.

There was consensus (92%) that age and occupational group should be considered when interpreting results of monofilament testing, and universal agreement that, given the paucity of normative data for Semmes Weinstein monofilament perception (SWM) perception in occupational groups, the 0.2 g-f cut off of normality should not automatically be increased for manual workers; however where finger tips are clearly thickened and the distribution of loss of sensory perception is symmetrical, this could be reflected in the interpretation of the SWM results.

There was consensus regarding the statement that those with peripheral neuropathy/neurological symptoms similar to neurological hand-arm vibration syndrome (HAVS) and wishing to work where exposed to hand-transmitted vibration (HTV), should be advised of the possible risks of further neurological loss in hands and fingers due to HTV, and should have a health surveillance assessment initially every 6 months for first two years by a clinician trained in detecting and diagnosing HAVS (if no evidence of progressive neurological deficit in the first two years, annual health surveillance should be considered if working with HTV). There was a range of opinion, failing to reach consensus regarding those with diabetes mellitus, who are at higher risk of CTS, and whether or not they should have quantitative sensory testing at baseline (before exposure to HTV) and then at regular intervals if working with HTV. There was 100% agreement that employees with diabetes mellitus, should not be excluded from exposure to HTV in order to mitigate legal risks for an employer associated with the diagnosis of a late stage of neurological HAVS.

There was a lack of consensus regarding the need for nerve conduction studies before confirming a diagnosis of CTS, whether cases of CTS should be restricted from using hand held vibratory tools until investigation and treatment is complete and whether cases of a recurrence of CTS should be permanently restricted from vibration exposure.

In respect of health surveillance, there was universal agreement that that following a new diagnosis of Stage 2 HAVS, frequency of Tier 4 (physician) assessment should be increased to 6 monthly, until there is no progression in symptoms. Where there has been a 2 year period in which there has been no symptom progression, assessment can revert back to a yearly, Tier 3 (occupational health adviser) or 4 (physician). There was unanimous disagreement with this statement (100% from 12 respondents) that Tier 5 testing (quantitative testing including thermal aesthesiometry and vibrotactile threshold measurement) was required for all cases of HAVS. The first round of the study elicited no overall consensus as to whether reduced sensory perception in HAVS can be staged by using only one quantitative sensory test (monofilaments) or whether quantitative sensory testing (QST) may play a useful role in refining a sensorineural grading of 2sn into "early" or "late", although in the second round there was agreement.

In respect of Dupuytren's disease, there was 100% agreement that employees with DD should not necessarily be restricted from vibration exposure at time of initial diagnosis regardless of severity or functional impairment. There was consensus (82%) that cases of DD should have enhanced health surveillance/periodic observations (e.g., every 6–12 months) to determine the onset of contracture and the need for referral and (91%) that restricting work with vibrating tools should be considered when functional impairment is such that it affects their ability to do work tasks or causing risk to others.

5. Conclusions

This Delphi review has provided a series of statements agreed by members of the UK Society of Occupational Medicine HAVS special interest group. It is anticipated that this will assist colleagues in the management of some of the difficult issues arising from vibration related disease in the workplace.

Author Contributions: Conceptualization, R.C.; methodology, R.C., D.A., H.F., C.G., R.H.-S., D.H., E.K. and I.L.; validation, D.A., H.F., C.G., R.H.-S., D.H., E.K. and I.L.; investigation, R.C., D.A., H.F., C.G., R.H.-S., D.H., E.K. and I.L.; writing—original draft preparation, review and editing, R.C., D.A., H.F., C.G., R.H.-S., D.H., E.K. and I.L. All authors have read and agreed to the published version of the manuscript.

Funding: Funding towards attendance and presentation at the 15th International Conference on HAVS in Nancy June 2023 was provided from the Society of Occupational Medicine Golden Jubilee fund.

Institutional Review Board Statement: Not applicable.

Informed Consent Statement: Not applicable.

Data Availability Statement: Not applicable.

Acknowledgments: We are grateful to the other members of the Delphi group for contributing to this study—Astrid Palmer, Isam Rustom, Rachel Sharp, Clare Piper, Anyanate Ephraim and Lachlan Mackay Brown. We acknowledge the support of the Society of Occupational Medicine, and in particular that of Nick Pahl, in supporting the work of the SIG in general and this project in particular.

Conflicts of Interest: The authors declare no conflict of interest. All authors are members of the SOM HAVS special interest group. The conclusions do not necessarily represent the views of the Society of Occupational Medicine or individual members thereof. IL is a member of the Industrial Injuries Advisory Council (IIAC). The conclusions do not necessarily represent the views of the IIAC or individual members thereof.

Reference

1. Barrett, D.; Heale, R. What are Delphi Studies? *Evid.-Based Nurs.* **2020**, *23*, 68–69. [CrossRef] [PubMed]

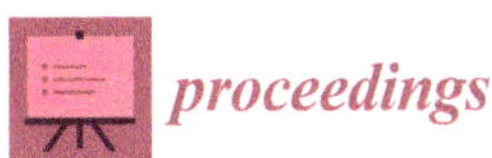

Proceeding Paper

Evaluation and Damping of High-Frequency Vibrations on a Tightening Tool [†]

Oscar Lundin and Romain Haettel *

Atlas Copco Industrial Technique AB, 131 34 Nacka, Sweden; oscar.lundin@atlascopco.com
* Correspondence: romain.haettel@atlascopco.com
† Presented at the 15th International Conference on Hand-Arm Vibration, Nancy, France, 6–9 June 2023.

Abstract: Impulse tightening tools can generate high-frequency vibrations that are not taken into account in currently applicable standards. In various publications, it has been suggested that those high-frequency vibrations may cause health issues. In the present study, high-frequency vibrations produced by an impulse nutrunner are evaluated and used to assess the potential damping effect provided by a thin layer of soft rubber.

Keywords: hand-arm vibrations; high-frequency material; damping material

1. Introduction

Hand-transmitted vibrations from power tools are associated with injuries and diseases labeled as hand-arm vibration syndrome (HAVS) [1,2]. In many countries, concerned authorities have thus issued regulations to protect users of vibrating tools [3]. For example, within the European Union, power tool suppliers are required to provide information on vibration emissions that reach or exceed a level of 2.5 m/s^2 measured and reported in accordance with several specific standards [4,5]. While the current methods for vibration declaration solely taking into account frequencies below 1250 Hz [2,5] have been useful in reducing the number of injuries, it is suggested in various studies [6–8] that vibrations at frequencies above 1250 Hz may cause nerve damages to power tool users. Consequently, there may be new requirements emerging from authorities and companies concerning the evaluation and the reduction of high-frequency vibrations.

The present paper deals with an experimental study of high-frequency vibrations measured on a hydraulic impulse nutrunner. In the study, the signals recorded on the nutrunner handle are used to assess the damping effects of a soft rubber layer on high-frequency vibrations.

2. Background and Procedure

In all industries in which productivity is essential, handheld power tools are utilized to perform tightening operations. In particular, hydraulic impulse nutrunners are widely used for assembly work due to rather high productivity and acceptable accuracy. Moreover, this power tool type presents obvious benefits in terms of ergonomics since they generate no significant reaction forces and have low vibration levels according to ISO 28927-2. In hydraulic impulse nutrunners, the torque is actually built up by impacts on a pulse unit containing oil that acts as a damping cushion. Nevertheless, the impacts may imply significant accelerations at high frequencies. Therefore, to investigate high-frequency vibrations and a potential damping method, measurements were performed on a hydraulic impulse nutrunner manufactured by Atlas Copco and designated EP7 PTI55.

The measurement procedure was conducted according to the general guidelines described in ISO 28927-2. However, the procedure was slightly simplified by using one machine run by only two operators. In addition, the time signals that were obtained from

Citation: Lundin, O.; Haettel, R. Evaluation and Damping of High-Frequency Vibrations on a Tightening Tool. *Proceedings* **2023**, *86*, 18. https://doi.org/10.3390/proceedings2023086018

Academic Editors: Christophe Noël and Jacques Chatillon

Published: 12 April 2023

the triaxial accelerometer mounted on the tool handle by using double-sided tape and a cable tie were acquired with a sampled frequency of 65,536 Hz. The recorded data were then analyzed by applying a low-pass filter at 10 kHz in order to conform to the useable frequency range of the accelerometer. For the measurements, the nutrunner was operated in the appropriate brake device at the maximum rated torque, as shown in Figure 1. Each of the two operators performed five runs of 10 s initially on the nutrunner with a regular handle (new tool from factory) and then on the same machine with a damped handle. The regular handle for this type of nutrunner is made of an aluminum core covered by hard rubber. The dampened handle consists of a 2 mm layer of soft rubber glued to the aluminum core and then covered by the hard rubber.

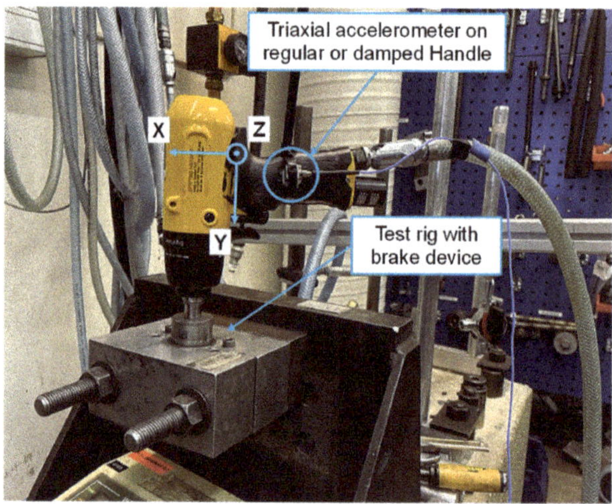

Figure 1. The nutrunner positioned on the brake device that was used for the vibration measurements that were performed with a triaxial accelerometer mounted on the handle.

The data provided by the triaxial accelerometer were used to evaluate the declared vibration emission value a_{hw} for the nutrunner. In addition, the Vibration Peak Magnitude (VPM) was calculated by using the relationship:

$$\text{VPM} = \sqrt{\frac{\sum a^{2+2k}}{\sum a^{2k}}} \qquad (1)$$

where the parameter k was set to two.

According to a report on signal processing [9], the value produced by the VPM calculation can be used to quantify the high-frequency content of the vibrations generated, for example, by repeated shocks. In the present study, the VPM value is mainly used to characterize the effect of vibration damping at high frequencies.

3. Measurement Results

In Figure 2, the time signals of the accelerations recorded by the triaxial accelerometer are shown in the X-, Y-, and Z-directions for two runs, where one run was performed on the regular handle and the other run on the damped handle. The signal amplitudes can vary significantly during a triggering sequence. Nevertheless, the signal amplitudes are clearly reduced by using the soft layer of damping material, especially in Y- and Z-directions.

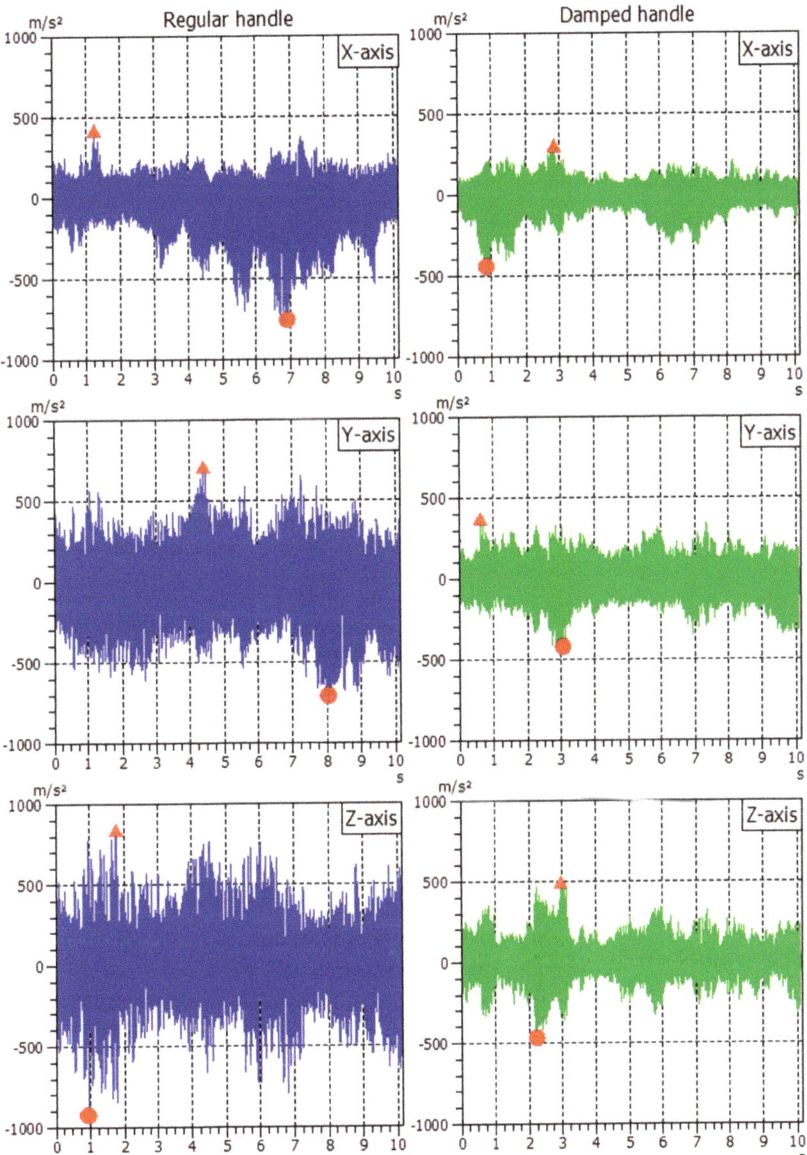

Figure 2. Time signals for two runs, one run performed on the regular handle (plots on the left) and the other run on the damped handle (plots on the right). Measurement direction and max-min values are indicated in each graph (red triangles for max values and red circles for min values).

The acceleration spectra obtained by averaging the vibration data from the ten runs are presented for both tool configurations in 1/24th octave bands in X-, Y-, and Z-directions, as shown in Figure 3. A large reduction of the acceleration amplitudes can be observed at frequencies above 1–2 kHz, whereas an increase in the acceleration levels can be viewed at the frequency range below 1 kHz, particularly in the Y-direction. This increase may be attributed to a resonance that is implied by the mounting of the accelerometer on the handle.

Proceeding Paper

Vibration Reduction of Pneumatic Rock Drill for Rock Face Stabilisation Sector [†]

Hans Lindell [1,*] , Thomas Clemm [2] and Snævar Leó Grétarsson [1]

[1] Department Manufacturing Processes, Research Institutes of Sweden(RISE), 431 53 Mölndal, Sweden; snaevar.gretarsson@ri.se

[2] STAMI—National Institute of Occupational Health in Norway, 0363 Oslo, Norway; thomas.clemm@mesta.no

* Correspondence: hans.lindell@ri.se

[†] Presented at the 15th International Conference on Hand-Arm Vibration, Nancy, France, 6–9 June 2023.

Abstract: Workers in the rock face stabilisation sector are exposed to high levels of vibration from pneumatic rock drills, which can lead to vibration injuries. The work situation is also ergonomically challenging since the work often is performed on steep cliffs with heavy equipment and a substantial degree of dust exposure. To reduce exposure to vibrations, the equipment has been redesigned, including the machine's handle, feeding hoist and the implementation of a reciprocating mass generating a counter force to reduce the vibrations. As a side project, a dust removal device was also developed. It was shown that vibration and dust exposure can be substantially reduced.

Keywords: rock drill; vibration reduction; HAVS; ATVA; vibration injury

Citation: Lindell, H.; Clemm, T.; Grétarsson, S.L. Vibration Reduction of Pneumatic Rock Drill for Rock Face Stabilisation Sector. *Proceedings* **2023**, *86*, 33. https://doi.org/10.3390/proceedings2023086033

Academic Editors: Christophe Noël and Jacques Chatillon

Published: 18 April 2023

1. Introduction

Rock drills expose operators for high vibration levels, which can lead to vibration injuries. These machines are frequently used for stabilising rock faces by drilling holes for securing bolts and steel nets to prevent rocks from falling on people and infrastructure. For operators in the rock face stabilising sector in Norway, pneumatic rock drills have been identified as a major source of vibrations [1,2]. Their work situation is also ergonomically challenging since the work is often performed on steep cliffs with heavy equipment and a substantial degree of dust exposure [3]. Drilling is also further complicated since the majority of holes are made horizontally into the rock for attaching fastening bolts and securing steel nets. The vertical force from the weight of the machine is supported by climbing ropes, and a horizontal feeding force of approximately 200 N to the drill is produced by a lever hoist attached to a drill and via a steel chain to a bolt in the rock. A common rock drill used for this work is Montabert T18. The operator is subjected to high levels of vibration in both hands, while operating the hoist and controlling the drill with the handle. During the ongoing project, improvements were made to the drill and supporting equipment to reduce the exposure to vibrations. The drilling setup can be seen in Figure 1. An important objective in this study is to show that there is a substantial potential for improvements on these type of machines for encouraging increased demands from operators and redesign activities from manufacturers.

Figure 1. Laboratory setup to recreate drilling conditions with current equipment when operators are horizontally drilling for the insertion of rock bolts. Prior to drilling the main hole, a smaller (8–10 mm), expanding bolt with an eye nut is placed to anchor the hoist chain. A lever hoist is used to control the feed force during drilling.

2. Method

Initial measurements were performed in a laboratory and a granite quarry to establish a baseline for the vibration level using standard equipment. Initial measurements of the tool showed two problem areas (see Table 1). The first one, as expected, was the high level of vibration of the handles. The second one was the high vibration level of the handle of the hoist. During drilling, the operator must keep one hand on the lever hoist to increase the feed force every few seconds. The problem with the current hoist is that the chain used is almost inelastic, which results in lots of variations in the feed force, i.e., when the operator uses the hoist, the feeding force rises sharply, but as soon as the drill is moved a few mm into the rock, the feeding tension drops to be close to zero.

Table 1. Results from vibration measurements in the project. Hand–arm-weighted acceleration RMS (ISO 5349-1:2001).

Tool	Handle Vibration (m/s^2_{haw})	Hoist Lever Vibration (m/s^2_{haw})
Original handle and hoist chain	34.6–40.8	25.4–40.9
Isolated handle and spring on hoist chain	15.7	5.4
ATVA, isolated handle and spring on hoist chain	11.6	5.4

The concept was split into three parts: spring–damper mechanism for lever hoist, isolated handle and auto-tuning vibration absorber (ATVA) on the machine to counteract the force from the impact mechanism. Since the overall objective of the project is to improve the working environment, a device that uses exhaust air for dust removal was also developed and is included in the new concept.

2.1. Spring–Damper Mechanism for Lever Hoist

A spring was added between the lever hoist and the machine. The spring has two functions. The first one is to limit the variation in the feed force. The spring can be stretched to its maximum length, and the operator does not have to use the hoist as often to avoid slack in the chain. The second function is to protect the hoist from the movement of the machine. By having the hoist fixed to the stationary rock wall and isolated from the machine, the vibration of the hoist lever can be reduced. The spring is contained in aluminium housing. The housing contains a small amount of oil for lubrication and prevents the spring from pinching the operator. The spring can be seen in Figures 2 and 3.

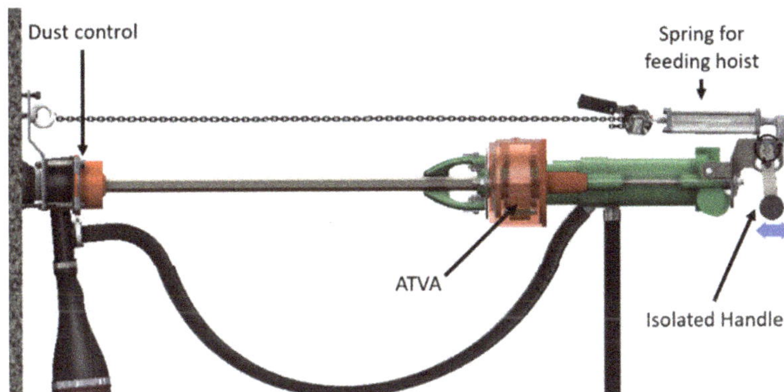

Figure 2. The combined concept for vibration reduction. A device for dust control was also developed during the project. The device redirects the dust that escapes the drill hole into a hose and releases it away from the operator.

Figure 3. The completed concept during laboratory testing.

2.2. Isolated Handle Solution

The vibration of the handle is determined by the drilling direction. The handle solution consists of a baseplate that is attached to the machine and handles that are connected to the machine with a lever arm. The lever arm is connected to the baseplate with two torsional isolators. This design was chosen to maximize vibration attenuation in the drilling direction without sacrificing controllability in the other two directions. Lever arm solutions have successfully been developed for similar rock drills in the past [4].

2.3. Auto-Tuning Vibration Absorber (ATVA)

Vibrations in the drilling direction were further reduced by adding an ATVA to the machine. The ATVA consists of 3 kg auxiliary mass with springs. The springs are not always in contact with the machine due to a gap, which gives the system a non-linear character. The springs' stiffness and the gap are optimized as described in [5]. The combined concept can be seen in Figure 2.

3. Results

A picture of the current prototype with the combined solutions, the hoist spring, isolated handle and ATVA, can be seen in Figure 3. The results from all the measurements in the project can be seen in Table 1.

The prototypes are currently being field tested by rock securers at Mesta AS in Norway. The prototype that is intended for use in the field can be seen in Figure 4.

Figure 4. The prototype without ATVA being tested in the field in Norway with the dust removal device.

4. Discussion and Conclusions

Initial measurements of the handle revealed extreme vibration levels of 34.6–40.8 m/s$^2_{haw}$. Combining an ATVA and isolated handle reduced the vibration level to 11.6 m/s$^2_{haw}$, which was measured during rock drilling. Implementing these solutions for current tools could reduce the operator's exposure to vibrations significantly. If these solutions were implemented by the manufacturer, as a part of a revision of the rock drill, the reduction could be even greater, and the weight of the machines could be reduced. The added weigh of the prototype with all the modification is 5–6 kg, which was compared with the original machine's weight of 22.3 kg. Although the operators want machines with a low weight, the benefits offset the added weight.

The hand–arm vibration level of the current hoist lever is 25–40.9 m/s$^2_{haw}$. By isolating the lever hoist from the machine, the vibration level of the lever was reduced to 5.4 m/s^2. This upgrade can be implemented without any modification of the rock drills; so, supplying all workers with this upgrade is simple. The results clearly show that there is a considerable potential for improving the very harsh working conditions for operators of these machines, where very few aspects have been improved over the last five decades.

Author Contributions: Conceptualization, methodology, writing and validation H.L., T.C. and S.L.G.; Investigation, formal analysis, data curation H.L. and S.L.G.; software and visualization, S.L.G.; project administration and funding acquisition H.L. and T.C. All authors have read and agreed to the published version of the manuscript.

Funding: This research was funded by Statens arbeidsmiljøinstitutt (STAMI) as a part of IA-Bransjeprogrammet for bygg og anlegg.

Institutional Review Board Statement: Not applicable.

Informed Consent Statement: Not applicable.

Data Availability Statement: Not applicable.

Conflicts of Interest: RISE hold patents for ATVA technology, and H. Lindell is the inventor.

References

1. Clemm, T. Hand-Arm Vibration Exposure in Rock Drill Workers: A Comparison between Measurements with Hand-Attached and Tool-Attached Accelerometers. *Ann. Work. Expo. Health* **2021**, *65*, 1123–1132. [CrossRef] [PubMed]
2. Clemm, T.; Lunde, L.K.; Ulvestad, B.; Færden, K.; Nordby, K.C. Exposure-response relationship between hand-arm vibration exposure and vibrotactile thresholds among rock drill operators: A 4-year cohort study. *Occup. Environ. Med.* **2022**, *79*, 775–781. [CrossRef] [PubMed]
3. Drilling and Rock bolting at Vøringsfossen. Fjellsikring. Available online: https://www.youtube.com/watch?v=OgsV05Ceprl&t=84s (accessed on 14 January 2023).
4. Marcotte, P. Development of antivibration handle for pneumatic jackleg rock drills. In *The Southern African Institute of Mining and Metallurgy Narrow Vein and Reef*; Southern African Institute of Mining and Metallurgy: Johannesburg, South Africa, 2008.
5. Lindell, H.; Berbyuk, V.; Josefsson, M.; Grétarsson, S.L. Nonlinear Dynamic Absorber to Reduce Vibration in Hand-held Impact Machines. In Proceedings of the International Conference on Engineering Vibration, Ljubljana, Slovenia, 7–10 September 2015.

Proceeding Paper

Study on Approaches for Reducing the Vibration Exposure of Hand-Helded Golf Club Heads Grinding [†]

Bin Xiao *, Yongjian Jiang, Wei Wen, Jianyu Guo, Maosheng Yan, Hansheng Lin and Shijie Hu *

Guangdong Province Hospital for Occupational Disease Prevention and Treatment, Guangdong Provincial Key Laboratory of Occupational Disease Prevention and Treatment (2017B030314152), Guangzhou 510300, China; jyj308978095@163.com (Y.J.); wenwcom@126.com (W.W.); gjy9306@outlook.com (J.G.); yanmsh@126.com (M.Y.); lhs1217@aliyun.com (H.L.)
* Correspondence: binny811@163.com (B.X.); hushijie@163.com (S.H.);
  Tel.: +86-020-89266013 (B.X.); +86-020-34063101 (S.H.)
† Presented at the 15th International Conference on Hand-Arm Vibration, Nancy, France, 6–9 June 2023.

Abstract: To control vibration-induced white finger among workers performing the fine grinding of golf club heads, in this study, the influence of different factors on vibration acceleration were verified by adjusting the eccentric mass of different dynamic balance wheels, adjusting the angle of driving wheels and passive wheels, increasing the number of rubber cushions, and using new and old sand belts. This study determined that the eccentric mass of the dynamic balance wheel, the number of rubber pads, and the newness of the sand belt are significant factors affecting the vibration of workers. The vibration hazard posed to workers can be reduced by correcting the dynamic balance of the front wheel, increasing the amount of rubber pads, and replacing the type of sand belt with a newer one.

Keywords: hand-arm vibration; hand-held workpiece vibration; vibration-induced white finger

1. Introduction

Hand-held workpieces grinding operation is a common hand-arm vibration operation. It is widely used in labor-intensive industries, such as hardware products, sports equipment manufacturing, and household appliance manufacturing. It is an essential link in the production process of enterprises and has a high risk of causing occupational diseases. The typical representative is a hand-held golf ball head polishing operation [1,2]. Previous researchers have reported that up to 30% of workers who perform the grinding of hand-held golf club heads suffered from hand-arm vibration disease [3]. The aims of this study are to explore the effects of various approaches for reducing the vibration exposure of hand-held golf club head grinding by comparing the difference in hand-transmitted vibration acceleration for three typical hand-held golf ball head polishing measures and to provide a basis for the prevention and control strategy of hand-held workpieces polishing operation.

2. Methods

This study consists of 4 experiments. Experiment (1) involves adjusting the balance of the driving wheel of the grinding equipment and analyzing the hand-transmitted vibration acceleration when the grinding driving wheel, weighted at 3.9 kg with different eccentric mass (3 g, 10 g, 15 g, 22 g, 32 g), rotates at 2100 RPM. Experiment (2) involves adjusting the relative position of the driving wheel (front wheel) and the passive wheel (rear wheel) of the grinding equipment and analyzing the hand-transmitted vibration acceleration under different situations wherein the angle between the front and rear wheels is 0 degrees and 5 degrees. Experiment (3) involves adjusting the damping cushion of the grinding equipment and analyzing the vibration acceleration when a single or double damping cushion is placed between the machine stand and the ground. Experiment (4) consists of analyzing and comparing the vibration acceleration using two kinds of sanding belts (WY60#, 3M50#).

For our experiments, we used an XL-168PA belt grinding machine with good performance (Hongweishun Corporation, Zhongshan, China), five kinds of driving wheels with different eccentric qualities (3 g, 10 g, 15 g, 22 g, 32 g), two kinds of rubber pads, two kinds of grind belts (WY60# 3M50#), and lots of golf club heads. An SV106 human vibration analyzer (SVANTEK, Warszawa, Poland) and a SV105B three-axis hand-transmitted vibration sensor (Poland SVANTEK) were used to measure the frequency-weighted accelerations in three orthogonal directions (x, y, and z), according to ISO 5349-1(2001) [4] and the previous research conducted by our research group [1]. The belt grinding machine is a type of classic bench grinding machine composed of a machine stand, motor, rotating shaft, driving wheel, driven wheel, sand belt, safety baffle, etc. The sand belt is fixed to a driving wheel and a driven wheel before operation. After the switch is turned on, the motor powers the driving wheel, sand belt, and driven wheel to rotate through the rotating shaft. The workpiece surface is polished through the rapid rotation of the sand belt.

Several experienced workers participated in this study. The testing grinders were randomly selected from a cohort of experienced grinders who had continuously worked on polishing jobs for more than five years. During experiments (1,2,3), ten club heads were polished. For the purpose of experiment(4), the test continued until the belt could not be polished.

SPSS 25.0 software was used for statistical analysis. A general linear model for the analysis of variance (ANOVA) was used to determine the effectual significance of our approaches. Whenever applicable, independent sample *t*-tests were performed to examine the significance of the vibration differences. Differences were considered significant at the $p < 0.05$ level.

3. Results

Figure 1 shows that five workers performed the same grinding operation by adjusting the eccentric mass of the driving wheel. There is an identifiably significant difference in the hand-transmitted vibration acceleration with different values of eccentric mass for the driving wheel during grinding. The higher the eccentric mass, the greater the hand-transmitted vibration acceleration ($p < 0.05$).

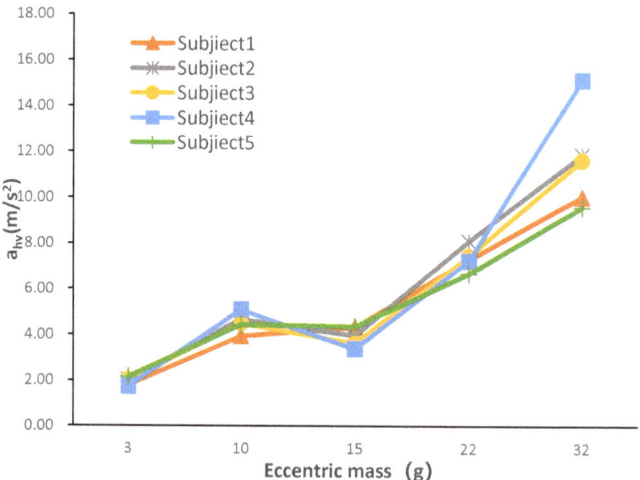

Figure 1. Comparison of results of frequency-weight accelerations in experiment (1), with five kinds of eccentric mass for the driving wheel (one subject but ten trials were performed for grinding with each eccentric mass of the driving wheel).

Table 1 shows that adjusting the angle of the driving wheel and the passive wheel led to no significant difference in the acceleration value of hand-transmitted vibration received by workers during grinding when the angle of driving wheel and passive wheel is 5° and parallel ($p = 0.628$).

Table 1. Results of the accelerations of hand-transmitted vibration received by workers during grinding when the angle of driving wheel and passive wheel is 5° and parallel.

The Angle of the Driving and Passive Wheels	A_{hv}, Mean ± STD (m/s^2)	t Value	p Value
5°	3.07 ± 0.36	0.491	0.628
<1°	3.01 ± 0.32		

Table 2 lists the results of putting rubber pads on the bottom of the machine. The average value of hand-transmitted vibration after adding double rubber pads is lower than adding a single rubber pad ($p < 0.05$).

Table 2. Comparison of frequency-weighted acceleration values of two kinds of pad.

Number of Rubber Pads	A_{hv}, Mean ± STD (m/s^2)	t Value	p Value
Single rubber pad	7.77 ± 0.56	−2.710	0.012
Double rubber pad	7.22 ± 0.48		

Figure 2 shows that the service life of the two different sand belts is less than that of the same type of a 3M sand belt. Among them, the WY60# sand belt can only grind approximately 19~23 golf club heads, while the 3M50# sand belt can grind approximately 33~37 heads. Moreover, The hand-transmitted vibration acceleration of the new belt (3M50#) was significantly lower than that of the old belt (WY 60#). As the number of grinding workpieces increased, the frequency-weighted acceleration values increased.

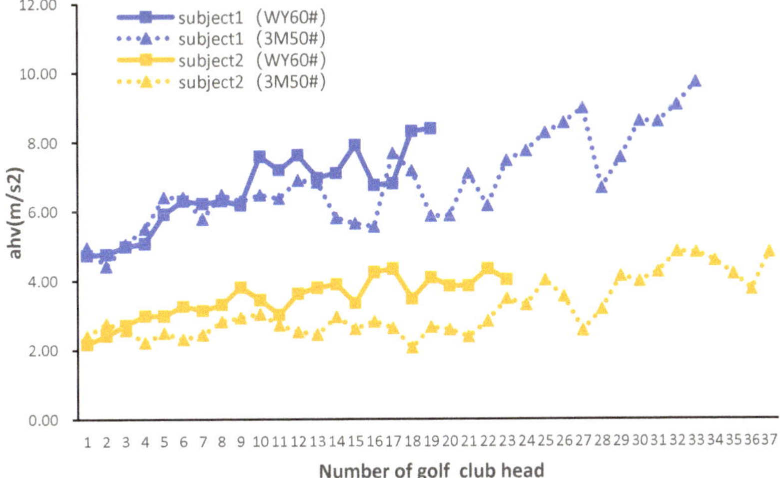

Figure 2. Comparison of the results measured in experiment 4 with two kinds of sand belts.

4. Discussion

We performed some early studies [1,5] to determine the vibration characteristics of golf club heads during the hand-held grinding process to find potential approaches for reducing vibration exposure. However, due to the limitations of our technological process, some possible measures proved difficult to realize, such as adjusting the motor speed, increasing the effective mass of the club head, etc. In this study, we tried to determine the influence of four kinds of practical measures that can be improved. Our research group compared the influence of different values of eccentric mass for the dynamic balance front wheel, the angle of driving wheel and passive wheel, the number of rubber cushions, and the new and old sand belts. The results showed that when the eccentric mass of the front wheel was increased, the vibration exposure level of test workers significantly improved. Adjusting the angle of the driving wheel and the passive wheel did not successfully reduce vibration levels. However, The vibration strength of the grinding workers can be noticeably reduced by adding more rubber pads. The hand-transmitted vibration acceleration of the new type sanding belt (3M 50#) was lower than that of the old type belt (WY 60#), and the service life of 3M in the same type belt is higher than that of the original belt. The above experiments show that, under normal operating conditions, correcting the balance of the front grinding wheel, installing double-layer rubber pads, and making use of the new model 3M sand belt all have a positive effect on reducing vibration hazards—therefore benefitting the workers. It should be noted that for most operations which involve potential exposure to hand-transmitted vibration, it is difficult to control the vibration exposure level and keep it within an acceptable range using a single method. The combination of engineering control and management control may be the best strategy for vibration prevention and control. In practical application, it should be comprehensively considered, along with the feasibility of implementation, cost-effectiveness, and its impact on production efficiency and workplace safety.

Author Contributions: Investigation, B.X., Y.J. and M.Y.; Methodology, B.X.; Project administration, S.H., B.X.; Supervision, B.X.; Measurement, J.G., Y.J. and H.L.; Data analysis, W.W., B.X.; Writing—original draft, B.X. and Y.J.; Writing—review & editing, B.X. All authors have read and agreed to the published version of the manuscript.

Funding: This research was funded by Guangdong Provincial Key Laboratory of Occupational Disease Prevention and Treatment (2017B030314152); Medical Scientifific Research Foundation of Guangdong Province, China (A2022062); Guangzhou Science and Technology Planning Project, grant number (201904010268).

Institutional Review Board Statement: The study was conducted in accordance with the Declaration of Helsinki, and approved by the Medical Ethics Committee of Guangdong Province Hospital for Occupational Disease Prevention and Treatment (protocol code GDHOD MEC 2021 No.047 and date of approval 23 November 2021) for studies involving humans.

Informed Consent Statement: Informed consent was obtained from all subjects involved in the study.

Data Availability Statement: The data presented in this study are available on request from the corresponding author. The data are not publicly available due to privacy.

Conflicts of Interest: The authors declare no conflict of interest.

References

1. Chen, Q.; Lin, H.; Xiao, B.; Welcome, D.E.; Lee, J.; Chen, G.; Tang, S.; Zhang, D.; Xu, G.; Yan, M.; et al. Vibration characteristics of golf club heads in their handheld grinding process and potential approaches for reducing the vibration exposure. *Int. J. Ind. Ergon.* **2017**, *62*, 27–41. [CrossRef] [PubMed]
2. Xiao, B.; Chen, Q.S.; Lin, H.S.; Wen, W.; Zhand, D.Y. Essential characteristics of hand-arm vibration in hand-held workpiece polishing. *China Occup. Med.* **2016**, *43*, 312–315.
3. Wei, N.; Lin, H.; Chen, T.; Xiao, B.; Ya, M.; Lang, L.; Yang, H.; Wu, S.; Boileau, P.É.; Chen, Q. The hand-arm vibration syndrome associated with the grinding of handheld workpieces in a subtropical environment. *Int. Arch. Occup. Environ. Health* **2021**, *94*, 773–781. [CrossRef] [PubMed]

4.   *ISO 5349-1:2001*; Mechanical Vibration—Measurement and Evaluation of Human Exposure to Hand-Transmitted Vibration—Part 1: General Requirements. International Organization for Standardization. Available online: https://www.iso.org/standard/3235 5.html (accessed on 12 April 2023).
5.   Lin, H.S. Relationship Between Vibration Exposure Level and Damage Effect of Hand-Held Workpieces Grinding Workers in a Guangdong Enterprise. Master's Thesis, Kunming Medical University, Kunming, China, May 2019.

 *proceedings*

Proceeding Paper

Evaluation and Damping of High-Frequency Vibrations on a Percussive Tool [†]

Romain Haettel * and **Oscar Lundin**

Atlas Copco Industrial Technique AB, 131 34 Nacka, Sweden; oscar.lundin@atlascopco.com
* Correspondence: romain.haettel@atlascopco.com
† Presented at the 15th International Conference on Hand-Arm Vibration, Nancy, France, 6–9 June 2023.

Abstract: Percussive tools can generate high-frequency vibrations that are not taken into account in currently applicable standards. In various publications, it has been suggested that those high-frequency vibrations may cause health issues. In the present study, high-frequency vibrations produced by a chipping hammer are evaluated and used to assess the potential damping effect provided by a thin layer of soft rubber.

Keywords: hand-arm vibrations; high frequency; damping material and percussive tools

Citation: Haettel, R.; Lundin, O. Evaluation and Damping of High-Frequency Vibrations on a Percussive Tool. *Proceedings* **2023**, *86*, 19. https://doi.org/10.3390/proceedings2023086019

Academic Editors: Christophe Noël and Jacques Chatillon

Published: 12 April 2023

1. Introduction

A low vibration level is, nowadays, an important feature for all hand-held power tools to comply with customer requirements and working environment regulations. In particular, vibration levels generated by percussive tools may cause not only discomfort but also health issues collectively designated as Hand-Arm Vibration Syndrome (HAVS) [1,2]. It is, therefore, crucial for operators performing repetitive tasks to use efficient tools emitting as low levels of vibration as possible [3]. For example, power tool suppliers are required within the European Union to provide information on vibration emissions that reach or exceed a level of 2.5 m/s^2 measured and reported in accordance with several specific standards [4,5]. Whereas the current methods for vibration declaration take into account solely frequencies below 1250 Hz, it has been suggested in various studies that vibrations in a frequency range above 1250 Hz may cause nerve damage to power tool users [6–8]. As a result, there might be new requirements emerging from authorities and companies concerning the evaluation of and the reduction in high-frequency vibrations.

The current article presents an experimental study of high-frequency vibrations measured on a chipping hammer. In the study, the signals recorded on the chipping hammer are used to assess the damping effects of a soft rubber layer on high-frequency vibrations.

2. Background and Procedure

Percussive tools for chipping and scaling are widely used for material removal tasks in foundries, metal workshops and shipyards. In those industries that require high productivity in rough working environments, chipping hammers shall offer robustness, sufficient percussive power and appropriate ergonomic features to protect the user from health issues, especially due to hand-arm vibrations. In fact, the oscillating forces that drive the piston in the percussive mechanism of a chipping hammer may generate high vibration levels in the tool handle. Therefore, certain models of chipping hammer have a damping system to reduce the vibrations transmitted to the handle. The vibration reduction can be achieved, for example, by using a soft spring that isolates the handle from the percussive mechanism. The chipping hammers with this type of damping system have relatively low vibration levels according to ISO 28927-10. However, the impacts produced in the percussive mechanism may still imply significant accelerations at high frequencies in the handle. Therefore, to investigate high-frequency vibrations and a potential damping method, measurements were performed on a damped chipping hammer manufactured by Atlas Copco and designated as RRF31.

The measurement procedure was conducted according to the general guidelines described in ISO 28927-10. However, the procedure was simplified by using one machine run by only two operators. In addition, the vibration signals were obtained from three accelerometers mounted with mechanical filters on a block attached to the tool handle by using double-sided tape and a cable tie. The time signals were acquired with a sampling frequency of 65,536 Hz. The recorded data were then analysed by applying a low-pass filter at 20 kHz. Moreover, the mechanical filters used to protect the accelerometers from potential overloads provided an additional low-pas filter at 10 kHz. For the measurements, the chipping hammer was operated in a steel ball energy absorber designated as Dynaload, as shown in Figure 1. Each of the two operators performed five runs of 10 s initially on the chipping hammer with the regular handle (new tool from factory) and then on the same machine with the handle partly covered with a 2 mm layer of soft rubber. The regular handle is made of aluminium with a painted surface.

The data provided by the three accelerometers were used to evaluate the declared vibration emission value a_{hw} for the chipping hammer. In addition, the Vibration Peak Magnitude (VPM) was calculated by using the relationship

$$\text{VPM} = \sqrt{\frac{\sum a^{2+2k}}{\sum a^{2k}}} \tag{1}$$

where the parameter k was set to 2.

According to a report on signal processing [9], the value produced by the VPM calculation can be used to quantify the high-frequency content of the vibrations generated, for example, by repeated shocks. In the present study, the VPM value is mainly used to characterize the effect of vibration damping at high frequencies.

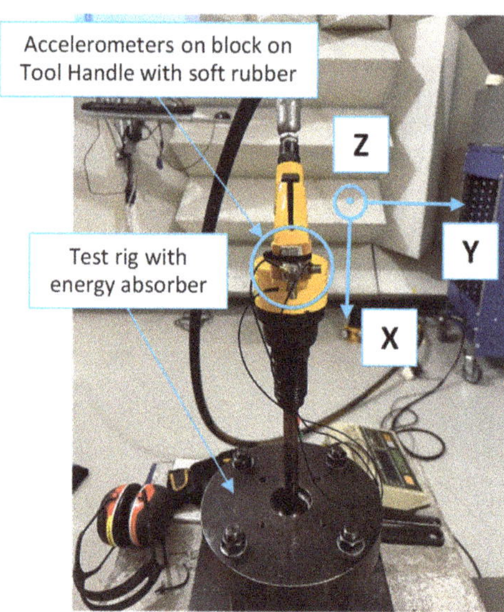

Figure 1. The chipping hammer positioned in the energy absorber used for the vibration measurements that were performed with three accelerometers mounted on the handle.

3. Measurement Results

In Figure 2, the time signals of the accelerations recorded by the three accelerometers are shown in the X-, Y- and Z-directions for two runs, where one run was performed on the regular handle and the other run on the damped handle. The signal amplitudes can vary significantly during a triggering sequence. Nevertheless, the signal amplitudes are clearly reduced by using the soft layer of damping material in the X-, Y- and Z-directions. The acceleration spectra obtained by averaging the vibration data from the 10 runs are presented for both tool configurations in 1/24th octave bands in X-, Y- and Z-directions, as shown in Figure 3. A large reduction in the acceleration amplitudes can be observed at high frequencies above 1 kHz, and there are no significant variations in the acceleration levels below 1 kHz.

In Table 1, the VPM and a_{hw} values obtained for both tool configurations are given in the X-, Y- and Z-directions and for the resulting vector norm. The soft rubber layer used on the damped handle provides a substantial decrease in the VPM value and, on the contrary, it has no effect on the vibration declaration value a_{hw}.

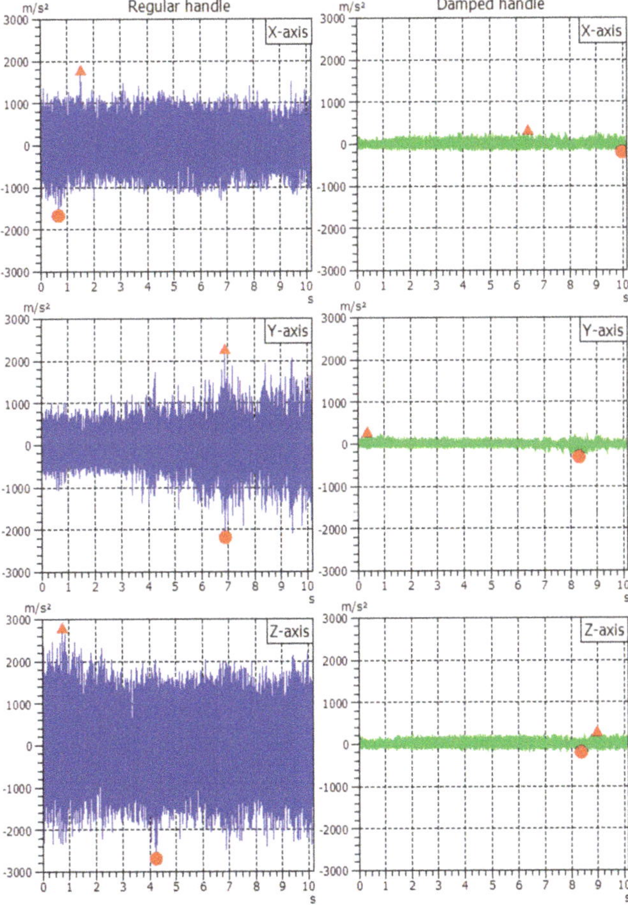

Figure 2. Time signals for two runs, one run performed on the regular handle (plots on the left) and the other run on the damped handle (plots on the right). Measurement direction and max–min values are indicated in each graph (red triangles for max values and red circles for min values).

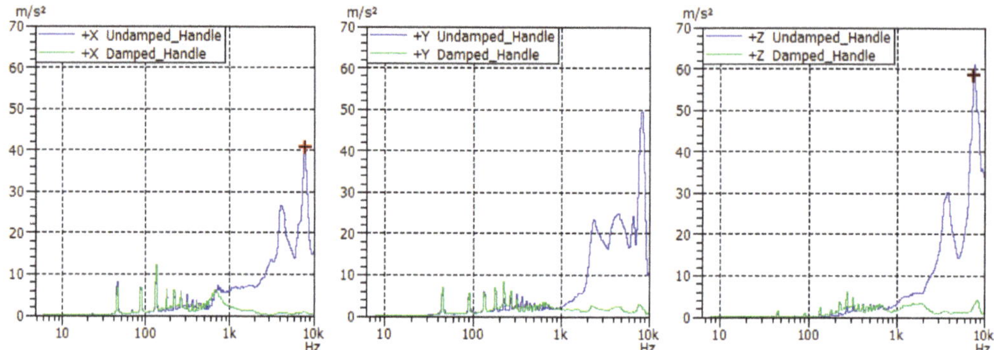

Figure 3. Acceleration spectra given in 1/24th octave bands for both tool configurations in X-, Y- and Z-directions.

Table 1. VPM and a_{hw} values are provided for both tool configurations with standard deviation and uncertainty factor, respectively. All values are given in m/s^2 and rounded to one decimal.

Tool	Regular Handle		Damped Handle	
	VPM (Std. Deviation)	a_{hw} (Std. Deviation)	VPM (Std. Deviation)	a_{hw} (Std. Deviation)
X:	859.8 (32.6)	4.2 (0.6)	156.8 (59.2)	3.8 (0.4)
Y:	1332.4 (311.3)	3.1 (0.6)	136.0 (25.1)	3.6 (0.7)
Z:	1585.2 (92.6)	0.8 (0.4)	112.4 (13.7)	1.1 (0.2)
Norm (X,Y,Z):	2260.5 (155,2)	5.3 (0.7)	242.8 (33.1)	5.3 (0.7)

4. Discussion and Conclusions

The damping effects of a thin layer of soft rubber on high-frequency vibrations are demonstrated by using time signals, acceleration spectra and VPM values. The soft rubber has no impact on the declared vibration emission value measured at 5.3 m/s^2. The official declaration value for the tool is 5 m/s^2 with an uncertainty K = 1.6 m/s^2. Furthermore, the damping effects at high frequencies, as shown in Figures 2 and 3, are clearly indicated by the decrease in the VPM values. This tends to prove that the VPM value can be used as a suitable parameter to quantify high-frequency vibrations. The use of a thin layer of soft rubber on a chipping hammer handle gives a substantial damping effect on high-frequency vibrations and provides extra insulation against the cold surfaces generated by the pneumatic tool. For further development, a suitable damped handle has to be designed and the material durability has to be investigated in industrial applications.

Author Contributions: R.H. and O.L. contributed equally to this article. All authors have read and agreed to the published version of the manuscript.

Funding: This research was funded by Atlas Copco Industrial Technique AB.

Institutional Review Board Statement: Not applicable.

Informed Consent Statement: Not applicable.

Data Availability Statement: The data are available on request if authorized by company policy.

Conflicts of Interest: There is a potential competing interest as the authors are employed by the tool manufacturer Atlas Copco Industrial Technique AB.

References

1. Griffin, M.J. *Handbook of Human Vibration*; Academic Press: London, UK, 1990.
2. *ISO 5349-1*; Mechanical Vibration—Measurement and Evaluation of Human Exposure to Hand-Transmitted Vibration—Part 1: General Requirements. International Organization for Standardization: Geneva, Switzerland, 2001.
3. *Directive 2002-44-EC*; The European Parliament and of the Council of 25 June 2002 on the Minimum Health and Safety Requirements Regarding the Exposure of Workers to the Risks Arising from Physical Agents (Vibration). The European Parliament and of the Council: Brussels, Belgium, 2002.
4. *Directive 2006-42-EC*; The European Parliament and of the Council of 17 May 2006 on Machinery. The European Parliament and of the Council: Brussels, Belgium, 2006.
5. *ISO 28927-10*; Hand-Held Portable Power Tools—Test Methods for Evaluation of Vibration Emission. International Organization for Standardization: Geneva, Switzerland, 2011.
6. *Hand-Arm Vibration: Exposure to Isolated and Repeated Shock Vibrations—Review of the International Expert Workshop 2015 in Beijing*; IFA Report; Deutsche Gesetzliche Unfallversicherung (DGUV): Berlin, Germany, 2017.
7. Lundström, R. Shock Vibration Caused by Percussive Hand Tools is an Underestimated Contributor to the Development of Vibration Injury. In Proceedings of the American Conference on Human Vibration (7th ACHV), Seattle, WA, USA, 13–15 June 2018.
8. Lindell, H.; Gerhardsson, L. Risk for VWF is underestimated in assembly industry using impact tools. In Proceedings of the Hand-Arm Vibration 14th International Conference, Bonn, Germany, 21–24 May 2019.
9. Johannisson, P.; Lindell, H. *Definition and Quantification of Shock-Impact-Transient Vibrations*; Technical Report; RISE: Mölndal, Sweden, 2022.

Proceeding Paper

Comparison of Anti-Vibration Glove Performances in the Laboratory and in the Field: Similarities and Differences [†]

Angelo Tirabasso, Raoul Di Giovanni, Pietro Nataletti and Enrico Marchetti *[iD]

INAIL, DIMEILA, Laboratory Physical Agents, 00078 Monte Porzio Catone, Italy; a.tirabasso@inail.it (A.T.); r.digiovanni@inail.it (R.D.G.); p.nataletti@inail.it (P.N.)
* Correspondence: e.marchetti@inail.it; Tel.: +39-0694181584
† Presented at the 15th International Conference on Hand-Arm Vibration, Nancy, France, 6–9 June 2023.

Abstract: Anti-vibration gloves undergo a certification procedure that is described in the EN ISO 10819:2013. This is a laboratory practice that, in this paper, is compared with a standardized field method. Two methods are implemented in the tests: a direct method and an experimental one. The results show that some transmissibility is above the limit suggested by the standard. The difference between the transmissibility measured in the field, using the direct transmissibility as a reference, and the values declared by the manufacturers is very small. In the case of the experimental transmissibility, two out of three gloves amplify vibrations instead of reducing them.

Keywords: anti-vibration gloves; transmissibility; vibration

Citation: Tirabasso, A.; Di Giovanni, R.; Nataletti, P.; Marchetti, E. Comparison of Anti-Vibration Glove Performances in the Laboratory and in the Field: Similarities and Differences. *Proceedings* **2023**, *86*, 9. https://doi.org/10.3390/proceedings2023086009

Academic Editors: Christophe Noël and Jacques Chatillon

Published: 10 April 2023

1. Introduction

The certification of anti-vibration gloves is performed following the directives of UNI EN ISO 10819:2013 [1,2]. This standard describes the standardized measurement protocol to be performed in the laboratory in order to certify an anti-vibration glove. When, on the other hand, measurements are taken in the field, the standardized measurement conditions are missing: the posture, grip strength, and vibration frequency of the tools, for example, are very different. The standard only recommends determining the transmissibility as a function of the frequency; then, the transmissibility varies depending on the tool used [3,4].

The purpose of this work is to compare the transmissibility of some anti-vibration gloves measured in the field with their certification in the laboratory, taking the manufacturer's certification data as the reference and comparing them with the data obtained from the measurements performed on a chainsaw.

2. Methods

The measurements were assessed on six male participants, experienced in using a chainsaw. The required task was to cut a pine trunk with a diameter of 15–20 cm perpendicular to the axis of the trunk in thin slices. Each operator first performed a bare hand measurement to obtain the transmissibility of the hand (T_0) and then a measurement with the anti-vibration glove to obtain the relative transmissibility (T_1). The following transmissibility was defined in line with the aforementioned UNI standard:

$$T_0 = a_{hv,h}/a_c,$$
$$T_1 = a_{hv,g}/a_c, \tag{1}$$

where $a_{hv,h}$, $a_{hv,g}$, and a_c represent the total acceleration value as defined by UNI EN ISO 5349-1 measured on the bare hand, on the gloved hand, and directly on the chainsaw handle, respectively.

The accelerations on the hand and on the handle were measured by means of two SEN026 PCB triaxial accelerometers, one inside the glove, inserted in a palmar adapter, and one fixed on the handle. A load cell was inserted inside the handheld adapter to measure the grip force. The operators were asked to comply with the real working conditions that arise during wood cutting operations, unlike in the laboratory where a standardized posture is prescribed [1].

The calculation of the transmissibility value was performed using the UNI EN ISO 10819:2013 model, i.e., 3 measurements for each of the 6 subjects, with and without gloves. Two methods were used to calculate the transmissibility, which we call the direct method and the experimental method:

1. Direct method: acceleration inside the glove/acceleration on the handle;
2. Experimental method: (gloved hand transmissibility/handle acceleration)/(bare hand transmissibility/handle acceleration) as prescribed by the standard (T_1/T_0).

The chainsaw transmissibility signal acquired in the field was divided into medium and high frequency for better comparability with the laboratory transmissibility described by the standard.

3. Results

The results were calculated using both the chainsaw handle and the bare hand as a reference and are described in Tables 1 and 2. The values declared by the manufacturer were also reported, where available. The values in bold are those that were beyond the limit value allowed by UNI EN ISO 10819 ($T_m \leq 0.9$; $T_h \leq 0.6$).

Table 1. The transmissibility calculated with the direct method for each subject.

Subject	Anti-Vibrating Glove					
Subject	Ansell 07-112		Ergodyne 9015		Impacto	
	T_m	T_h	T_m	T_h	T_m	T_h
1	0.88	**0.81**	0.88	0.55	0.88	**0.81**
2	0.74	**0.74**	0.70	**0.68**	0.74	**0.74**
3	0.74	**0.68**	0.46	0.52	0.85	**0.82**
4	0.61	0.52	0.50	0.52	0.61	0.57
5	0.55	0.43	0.46	0.36	0.66	0.57
6	0.81	0.44	0.77	0.40	0.34	0.34
Mean ± SD [1]	0.72 ± 0.12	0.60 ± 0.16	0.63 ± 0.18	0.50 ± 0.12	0.68 ± 0.20	**0.64 ± 0.18**
Declared	0.90	0.52	0.80	0.57	N.a. [2]	N.a.

[1] S.D. = Standard Deviation; [2] N.a. = Data not available.

Table 2. The transmissibility calculated with the experimental method for each subject.

Subject	Anti-Vibrating Glove					
Subject	Ansell 07-112		Ergodyne 9015		Impacto	
	T_m	T_h	T_m	T_h	T_m	T_h
1	0.79	**0.79**	0.59	**0.66**	0.82	**0.98**
2	0.82	**0.80**	0.80	**0.83**	0.84	**0.89**
3	**1.02**	**0.98**	0.64	0.44	**1.16**	**1.17**
4	0.85	**0.62**	0.62	0.51	0.77	**0.68**
5	0.73	**0.61**	0.62	0.51	0.88	**0.81**
6	0.80	0.45	0.76	0.40	0.34	0.35
Mean ± SD [1]	0.84 ± 0.10	**0.71 ± 0.18**	0.67 ± 0.09	0.56 ± 0.16	0.80 ± 0.27	**0.81 ± 0.28**
Declared	0.90	0.52	0.80	0.57	N.a. [2]	N.a.

[1] S.D. = Standard Deviation; [2] N.a. = Data not available.

4. Discussion

The difference between the transmissibility measured in the field, using the direct transmissibility as a reference, and the values declared by the manufacturers was very small, with some exceptions, and the standard deviation of the values measured in the field had good values. In the case of the experimental transmissibility, however, two out of the three gloves would not pass the certification because, for some subjects, they amplified rather than reduced the vibrations.

The transmissibility values obtained respectively with reference to the bare hand and with the values declared by the manufacturers were quite close to each other. This was even more evident by comparing the standard deviation obtained on several measurements and that obtained for each individual subject. These data were not reported for reasons of space.

The current data seemed to confirm the effectiveness of the certification protocol by comparing the measured data with those declared. It seems reasonable to assume, given the high standard deviation, that in the field it would be right to measure a larger number of subjects to account for the high variability. The next step to conclude the work will be to carry out laboratory measurements by reproducing the average signal obtained in the field on the chainsaw and adopting the indications of the UNI EN ISO 10819:2013 certification standard.

Author Contributions: Conceptualization, E.M. and A.T.; methodology, P.N. and R.D.G.; software, R.D.G.; validation, E.M. and A.T.; formal analysis, R.D.G.; resources, E.M.; data curation, R.D.G.; writing—original draft preparation, P.N.; writing—review and editing, E.M. and A.T.; supervision, E.M. All authors have read and agreed to the published version of the manuscript.

Funding: This research received no external funding.

Institutional Review Board Statement: The study was conducted in accordance with the Declaration of Helsinki, and approved by the Institutional Review Board of INAIL.

Informed Consent Statement: Informed consent was obtained from all subjects involved in the study.

Data Availability Statement: Not applicable.

Acknowledgments: We acknowledge the useful help provided by ASU Central Friuli.

Conflicts of Interest: The authors declare no conflict of interest.

References

1. *ISO 5349-1:2001*; Mechanical Vibration—Measurement and Evaluation of Human Exposure to Hand-Transmitted Vibration—Part 1: General Requirements. International Organization for Standardization: Geneva, Switzerland, 2001.
2. *ISO 10819:2013*; Mechanical Vibration and Shock—Hand-Arm Vibration—Measurement and Evaluation of the Vibration Transmissibility of Gloves at the Palm of the Hand. International Organization for Standardization: Geneva, Switzerland, 2013.
3. Pinto, I.; Stacchini, N.; Bovenzi, M.; Paddan, G.S.; Griffin, M.J. Protection Effectiveness of Anti-Vibration Gloves: Field Evaluation and Laboratory Performance Assessment; Appendix H4C to Final Report; EC Biomed II concerted action BMH4-CT98-3291. In Proceedings of the 9th International Conference on Hand-Arm Vibration, Nancy, France, 5–8 June 2001.
4. Nataletti, P.; Lenzuni, P.; Lunghi, A.; Pieroni, A.; Marchetti, E. Antivibration gloves effectiveness: In field and laboratory tests and proposal for a new standard. In Proceedings of the 10th International Conference on Hand-Arm Vibration, Las Vegas, NV, USA, 7–11 June 2004.

Proceeding Paper

Fingertip Model for Analysis of High-Frequency Vibrations [†]

Peter Ottosson *[iD], Hans Lindell [iD] and Snævar Leó Grétarson [iD]

RISE Research Institutes of Sweden, SE-431 53 Moelndal, Sweden; hans.lindell@ri.se (H.L.); snaevar.gretarsson@ri.se (S.L.G.)

* Correspondence: peter.ottosson@ri.se

† Presented at the 15th International Conference on Hand-Arm Vibration, Nancy, France, 6–9 June 2023.

Abstract: High-frequency shock-type vibrations (HFVs) from, e.g., impact wrenches with a frequency content mainly above 1250 Hz have long been suspected to cause a significant number of vibration injuries, HAVS. These vibrations are unregulated in the current standard for risk estimation, ISO 5349-1; thereby, the risk of injury is suspected to be underestimated. The objective of this study was to investigate the effects on finger tissue subjected to HFVs similar to those from impact wrenches by using a 2D finite element model of a fingertip. The model was validated through experiments. Using the input acceleration from the experiments, the model predicted high pressure variation and particular negative pressures at levels close to 0.1 MPa (1 Bar) or more, which are levels where cavitation in liquid can occur, with a detrimental effect on biological systems.

Keywords: high-frequency vibration; ultravibration; HAVS; vibration injury; impact wrench; shock vibration; numerical model; experimental validation; negative pressures; material properties

1. Introduction

High-frequency shock-type vibrations (HFVs) with an energy content mainly above 1250 Hz are likely to cause a significant number of vibration injuries [1], but there is an inadequate understanding of the injury mechanisms. There is also no standard assessing the risk associated with these vibrations. The current standard, ISO 5349-1, that all regulations and legislations are based on only covers the octave bands from 8 Hz to 1000 Hz. The objective of this study was to gain knowledge on the physical entities inside a finger at the local tissue level where injures are suspected to occur when subjected to HFVs, in order to develop mitigation measures. To enable detailed studies of the propagation of HFVs occurring in the use of, e.g., an impact wrench, a 2D finite element model of a fingertip was developed. The geometries of the constituents are schematic to resemble the fingertip morphology rather than being an exact representation of an existing finger.

2. Materials and Methods

2.1. Model

The model of a schematic fingertip is shown in Figure 1. This model includes the major components of the human skin, i.e., stratum corneum, epidermis, dermis and subcutaneous tissue. The geometric properties of the different skin layers and the overall dimensions of the finger were derived from findings published in [2,3]. Special attention is paid to the contour of the fingerprint where load introduction appears. The 3D contour was scanned with imaging confocal microscopy and parametrized. The dimensions were integrated into the simulation model. The model consists of 63,500 2D elements (plane strain) located in the global xy-plane. Element sizes range from 0.01 mm (the smallest) in the stratum corneum to 0.08 mm in the bone. A fully integrated 2D plane strain element formulation with four in-plane integration points was used throughout this work, counteracting any spurious deformations, i.e., hourglass modes. This continuum element enables the use of material models corresponding to 3D continuum elements.

Citation: Ottosson, P.; Lindell, H.; Grétarson, S.L. Fingertip Model for Analysis of High-Frequency Vibrations. *Proceedings* **2023**, *86*, 23. https://doi.org/10.3390/ proceedings2023086023

Academic Editors: Christophe Noël and Jacques Chatillon

Published: 13 April 2023

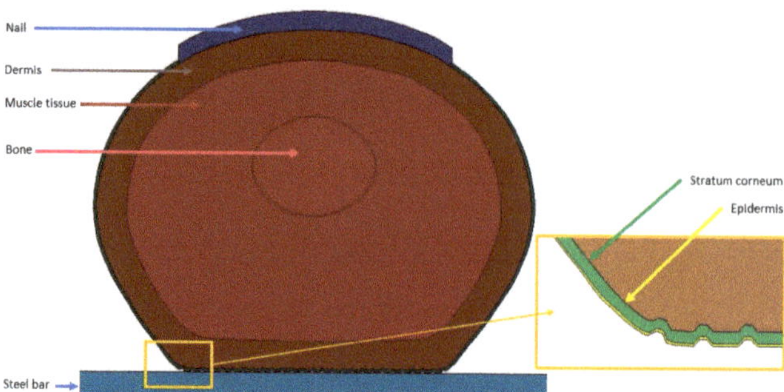

Figure 1. Two-dimensional fingertip model.

2.2. Material Data and Material Models

The difficulties in determining the material parameters for living tissue to be used in numerical modeling are notorious. Literature data on the mechanical properties of the skin layers (stratum corneum, epidermis, dermis and subcutaneous tissue) reveal differences in the order of magnitude, depending on the test setup, loading conditions, gender, age, location and environmental conditions [2–6]. Material data for the stratum corneum and epidermis stated in [2] were further refined through an experimental investigation of the finger and fingerprint distortion under compressive forces. These results were used to validate the finite element simulation model using an inverse optimization approach [7]. The material response of the skin layers is time- and history-dependent; therefore, a viscoelastic constitutive model, based on exponential stress relaxation functions, was used [8,9].

2.3. Experimental Data for the Modeling

The experimental setup is shown in Figure 2. It generates a vibration pulse with high repeatability and amplitude, similar to ordinary hand-held impact wrenches. From this experiment, the acceleration of the steel surface was extracted. After post-processing, this signal was used as the input for the FE model to avoid numerical disturbances. The response in the nail was measured up to 100 kHz using a laser Doppler vibrometer (LDV) that registered the resulting accelerations at the nail tip. The signal was sampled at 1 MHz.

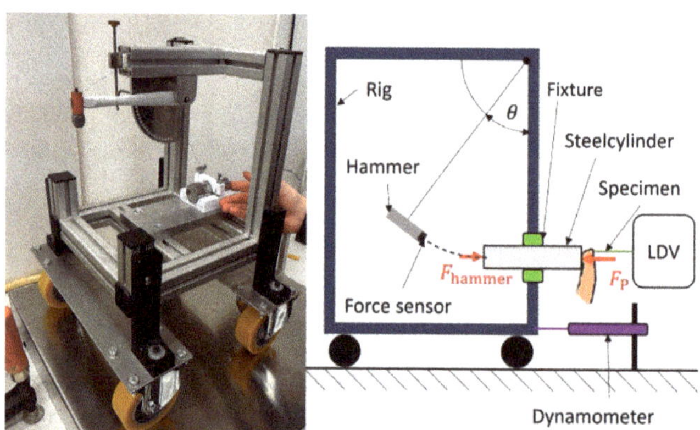

Figure 2. Photo of the experimental setup used and schematic of the experimental setup.

2.4. Numerical Aspects

The commercial software LS-Dyna was used for the finite element simulations. Two different numerical schemes were used, i.e., implicit and explicit time integration. In the experiments, a pre-load was applied; thus, a pre-load was required even in the numerical models. The static nature of this pre-loading made it more suitable to use implicit time integration for this phase. Once the pre-loading face was fulfilled, the switch between the numerical algorithms was carried out automatically, and the program entered the explicit time integration scheme. The high acceleration dynamics in the excitation required explicit time integration, where time stepping was continuously incremented, basing the time step on the smallest dimension of all the elements in the model.

3. Results

The measured response of the nail acceleration in the experiments was used to compare the experiment with the simulations using the model (Figure 3). The correlation for the first measured acceleration peaks, which are considered to be of utmost importance, was fairly good between the experiment and FE simulation, as can be seen in Figure 3. Then, the correspondence deteriorated, even though some peaks were later comparable.

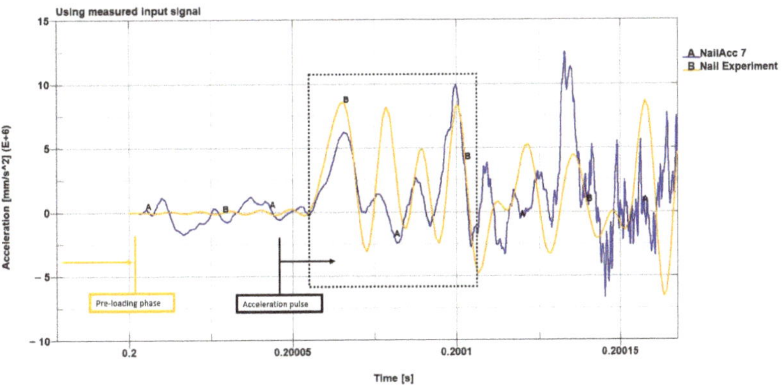

Figure 3. Measured acceleration on the nail surface (yellow) compared with the FE result (blue) from the corresponding point on the nail.

Pressure variation is considered vital for the consequences of shock-vibration-related tissue damage. Notably, negative pressures at levels close to 1 bar (equal to 0.1 MPa) or more were found in the stratum corneum and epidermis in these analyses (Figure 4).

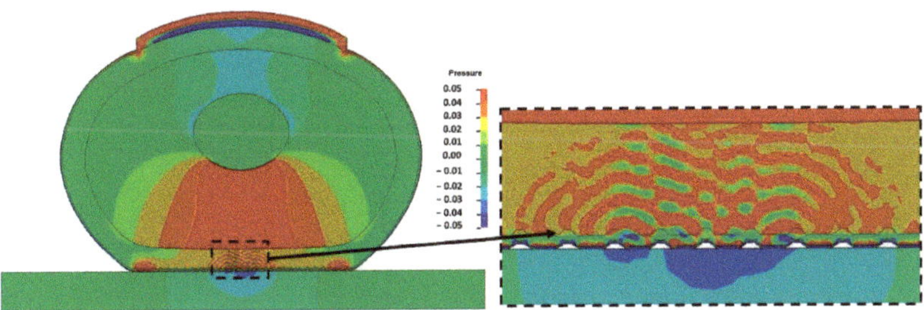

Figure 4. Pressure field in Mpa in the model immediately after the acceleration pulse hit the steel bar. N.B. Negative pressures are shown in green to blue colors.

4. Discussion

The first peak is of major importance, and there can be many reasons for the discrepancies between the experiment and simulation with respect to this; the material parameters will have an influence, and the values may show significant variation between individuals.

Numerical contact formulations can contribute to numerical noise. In fact, the switching from an implicit solution technique to a subsequent explicit analysis can cause oscillations that can be detected if the acceleration signal is applied with a latency time.

Another concern is that strain rates can be of significance when it comes to tissue damage. Currently, it is not known at what level of strain rate certain tissues are at risk. High strain rates might also suggest the use of more elaborate material models to study such effects accurately.

5. Conclusions

The 2D fingertip model can be pre-loaded and exposed to accelerations from experiments to yield the first-order correspondence with the experimentally measured output. Through this, it is assumed, but not proven, that the physical entities in the internal structures as a consequence of these mechanical loads correspond to the structures in a finger experienced in such an environment.

It would be invaluable to gain access to in situ measurements of the response in the fingertip to correlate the simulation results. Measurement of the pressure in the respective structural layers of a fingertip can provide results that can be used for the validation of the model. This can be achieved by measuring the pressure using, e.g., a catheter with a Fabry–Pérot etalon.

Data used in the model can be improved in various ways. The described uncertainties with respect to the material data and material models, as well as the vast variation in the human population, are one area of possible improvement.

The model itself can be improved in various ways, from elaborating the geometry or using a more morphologically correct configuration to utilizing a higher mesh resolution and more advanced numerical techniques. Pressures could be vital for the consequences of shock-vibration-related tissue damage. Negative pressures at levels close to 1 bar (equal to 0.1 MPa) or more were found in these analyses, which are levels where cavitation in liquid can occur, with a detrimental effect on biological systems. The results indicate the need for deeper knowledge of the physical processes at the tissue level in humans in order to prevent vibration injury.

Author Contributions: H.L. and S.L.G. conceived and planned the experiments. H.L. and S.L.G. carried out the experiments. P.O. planned and carried out the simulations. H.L. supervised the findings of this work. All authors have read and agreed to the published version of the manuscript.

Funding: This research was funded by DGUV Forschungsförderung, FP-415 (www.dguv.de).

Institutional Review Board Statement: The study was conducted in accordance with the Declaration of Helsinki, and approved by the Ethics Committee of Luebeck University (protocol code 20-099, April 2020).

Informed Consent Statement: Not applicable.

Data Availability Statement: Not applicable.

Conflicts of Interest: The funders had no role in the design of the study; in the collection, analyses or interpretation of data; in the writing of the manuscript; or in the decision to publish the results.

References

1. Dandanell, R.; Engström, K. Vibration from riveting tools in the frequency range 6 Hz-10 MHz and Raynaud's phenomenon. *Scand. J. Work Environ. Health* **1986**, *12*, 338–342. [CrossRef] [PubMed]
2. Vodlac, T.; Vidrih, Z.; Sustaric, P.; Rodic, T.; Wessberg, J.; Peric, D. Computational assessment of spiking activity during surface exploration: From finite element analysis to firing-rate. In *Haptics: Perception, Devices, Control and Applications*; Springer: Berlin/Heidelberg, Germany, 2016.
3. Geerlings, M. *In Vitro Mechanical Characterization of Human Skin Layers: Stratum Corneum, Epidermis and Hypodermis*; Technical University Eindhoven: Eindhoven, The Netherlands, 2006.
4. Leyva-Mendivil, F.; Page, A.; Bressloff, N. A mechanistic insight into the mechanical role of the stratum corneum during stretching and compression of the skin. *J. Mech. Behav. Biomed. Mater.* **2015**, *49*, 197–219. [CrossRef] [PubMed]

5.   Pal, S. *Design of Artificial Human Joints and Organs*; Springer: New York, NY, USA, 2014.

6.   Goss, S.A.; Johnston, R.L.; Dunn, F. *Comprehensive Compilation of Empirical Ultrasonic Properties of Mammalian Tissues*; Bioacoustics Research Laboratory, University of Illinois: Urbana-Champaign, IL, USA, 1978.

7.   Lindell, H.; Grétarsson, S.L.; Machens, M. High Frequency Shock Vibrations and implications of ISO 5349—Measurement of vibration, Simulating Pressure Propagation, Risk Assessment and Preventive Measures. In Proceedings of the Report from Workshop on Single Shocks at 13th International Conference on Hand-Arm Vibration, Bejing, China, 16 October 2015.

8.   Herrman, L.; Peterson, F. A numerical procedure for viscoelastic stress analysis. In Proceedings of the 7th Meeting of ICRPG Mechanical Behavior Working Group, Orlando, FL, USA, 14–15 November 1968.

9.   Krave, U. *Modelling of Diffuse Brain Injuries: Combining Methods to Study Possible Links between Transient Intracranial Pressure and Injury*; Doctoral thesis of Engineering; Chalmers University of Technology: Gothenburg, Sweden, 2010.

Proceeding Paper

Factoring Muscle Activation and Anisotropy in Modelling Hand-Transmitted Vibrations: A Preliminary Study [†]

Simon Vauthier [1,2], Christophe Noël [1,*], Nicla Settembre [3], Ha Hien Phuong Ngo [4], Jean-Luc Gennisson [4], Jérôme Chambert [2], Emmanuel Foltête [5] and Emmanuelle Jacquet [2]

[1] Electromagnetism, Vibration, Optics Laboratory, Institut national de recherche et de sécurité (INRS), 54519 Vandœuvre-lès-Nancy, France; simon.vauthier@inrs.fr

[2] Université de Franche-Comté, CNRS, Institut FEMTO-ST, F-25000 Besançon, France; jerome.chambert@univ-fcomte.fr (J.C.); emmanuelle.jacquet@univ-fcomte.fr (E.J.)

[3] Department of Vascular Surgery, Nancy University Hospital, University of Lorraine, 54500 Vandœuvre-lès-Nancy, France; nicla.settembre@univ-lorraine.fr

[4] BioMaps, INSERM, CEA, CNRS, Université Paris-Saclay, 91401 Orsay, France; ha-hien-phuong.ngo@universite-paris-saclay.fr (H.H.P.N.); jean-luc.gennisson@universite-paris-saclay.fr (J.-L.G.)

[5] SUPMICROTECH, CNRS, Institut FEMTO-ST, F-25000 Besançon, France; emmanuel.foltete@ens2m.fr

* Correspondence: christophe.noel@inrs.fr; Tel.: +33-3-83-50-21-12

[†] Presented at the 15th International Conference on Hand-Arm Vibration, Nancy, France, 6–9 June 2023.

Abstract: Pushing and gripping forces may contribute to Hand-Arm Vibration Syndrome but, thus far, have not been taken into account in vibratory dose assessment according to the current standards. To obtain a better understanding of the symptom onset, we developed a finite element model of the hand to replicate its vibratory behaviour in gripping and pushing actions. In a case study, Supersonic Shear Imaging measurements revealed the significant dependence of muscle stiffness and anisotropy on gripping. The use of these measurements in our model showed that muscle activation influences the driving-point mechanical impedance of the hand and local vibration propagation.

Keywords: vibration hazard; elastography; muscle activation; transversally isotropic; modelling

Citation: Vauthier, S.; Noël, C.; Settembre, N.; Ngo, H.H.P.; Gennisson, J.-L.; Chambert, J.; Foltête, E., Jacquet, E. Factoring Muscle Activation and Anisotropy in Modelling Hand-Transmitted Vibrations: A Preliminary Study. *Proceedings* **2023**, *86*, 12. https://doi.org/10.3390/proceedings2023086012

Academic Editor: Jacques Chatillon

Published: 10 April 2023

1. Introduction

In France, nearly 2.2 million workers are exposed to hand-transmitted vibration. Prolonged exposure to high-level vibration can lead to various disorders, known as Hand-Arm Vibration Syndrome [1]. In an attempt to reduce the health effects, the daily vibration dose received by workers is limited by law in Europe. However, dose assessment [2] has certain shortcomings. In particular, pushing and gripping forces exerted by the operator are not taken into account, although they significantly influence the hand's driving-point mechanical impedance (DPMI), which has been identified as a potential marker of vibration hazard [3]. To better understand the symptom onset, we investigated coupling force effects on vibration propagation in specific hand regions. Hence, we developed a finite element (FE) model for mimicking pushing and gripping actions to simulate their effects on the vibratory behaviour of the hand. The first step was to induce increased muscular stiffening stemming from muscle activation related to gripping. Supersonic Shear Imaging (SSI) was performed for local measurements of the shear elastic modulus in the hand muscles [4]. Thus, our approach consisted of measuring the stiffness of the first dorsal interosseous muscle (FDIM) of the hand as a function of gripping force. In addition, this technique allows for the measurement of the mechanical properties of muscles in directions parallel and perpendicular to their fibres. These measurements were then used to feed our FE model with muscle constitutive laws depending on both muscle activation and anisotropy. The aim of this paper is to demonstrate the feasibility of this approach in a case study and to quantify the influences of gripping-induced muscle disturbances on both DPMI and local vibratory transmissibility.

2. Materials and Methods

2.1. Measurement of Muscle Elasticity with Supersonic Shear Imagining

The shear elastic modulus was measured with an Aixplorer® ultrasonic scanner (Aix-en-Provence, France) in the SSI mode [4]. An experimental apparatus was set up to measure the shear elastic modulus of the FDIM of a volunteer gripping a handle instrumented with force sensors (Figure 1a). The probe was oriented beforehand using B-mode imaging either parallel or perpendicular to the orientation of the muscle fibres. The volunteer followed a given instruction by managing the grip level displayed on a screen. The protocol consisted of first measuring the maximum grip force solely to estimate the relative gripping forces. Next, the subject followed randomly chosen instructions for the application of gripping forces ranging from 0 to 30% with 5% increments. Both fibre directions were measured. The non-smoker volunteer was a 23-year-old male in good health. The shear elasticity modulus resulted in averaging data contained in a 5×5 mm^2 region-of-interest (ROI) extracted from raw elastography maps (Figure 2a,b).

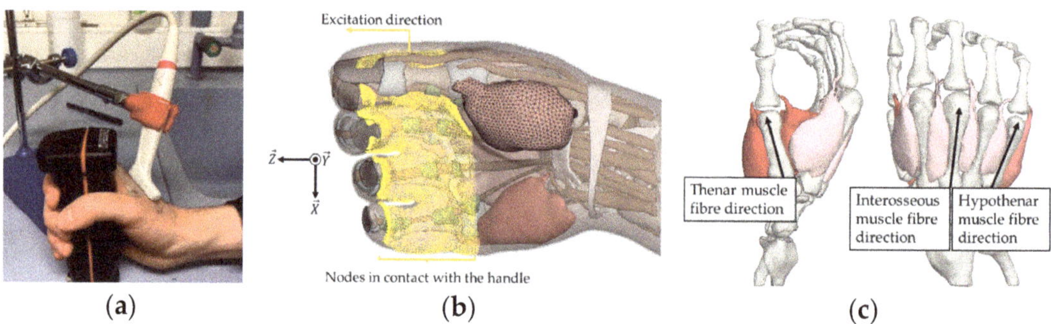

Figure 1. (**a**) Setup for measuring shear modulus with an ultrasonic probe while gripping an instrumented handle. (**b**) FE model of the hand. (**c**) Fibre direction for orthotropic constitutive laws.

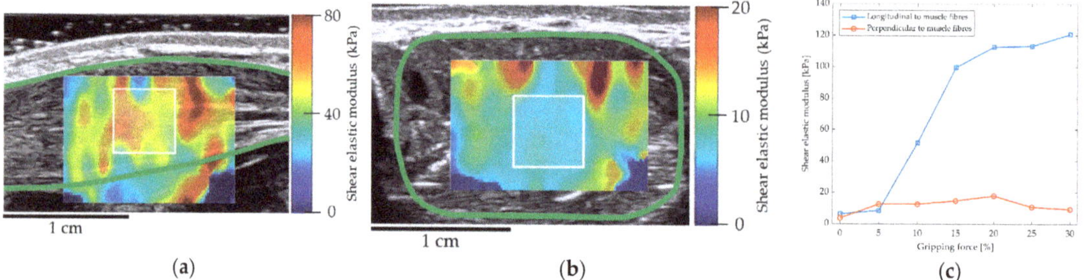

Figure 2. Examples of shear elastic modulus measurements obtained by SSI in the FDIM: (**a**) parallel and (**b**) perpendicular to fibres. The background represents the B-mode image with the muscle enclosed in green. The white square, which is 5 mm in length. indicates the ROI. (**c**) Average of longitudinal and perpendicular shear elastic moduli in the ROI as a function of gripping force.

2.2. FE Modelling of Hand-Transmitted Vibrations

Our FE hand model was built by segmenting MRI images of the hand of a 28-year-old male volunteer [5]. Most of the hand anatomical elements were included and meshed with tetrahedrons of around 1 mm (Figure 1b). The muscles were divided into three groups: interosseous, thenar and hypothenar. For each group, the muscle fibre direction matched the closest metacarpal bone direction (Figure 1c). The hand position corresponded to the grip on a handle without tightening. The initial stress and deformation fields were therefore taken to be zero. There were no boundary conditions placed on the wrist. The handle was excluded from the model, and the skin nodes in contact with it were fastened in all directions except for the direction of excitation (Figure 1b). The DPMI was calculated and compared to the standard ISO 10068 [6]. In addition, the vibratory transmissibility was computed in two areas: in the tissue under the median phalanx of the index (where an artery is likely to pass, not included in the model) and in the FDIM. Harmonic analyses were carried out by modal superposition over a range of 10–400 Hz. Modal damping was added to the system (from 17% to 2%, decreasing in frequency up to 200 Hz and remaining constant beyond). The calculations were performed using the FE software LS-Dyna$^{®}$ (Ansys, Canonsburg, PA, USA) by assuming linear elastic constitutive laws. The parameters were derived from [5], except those for the muscles, which were derived from elastography measurements. Three cases were computed: (i) isotropic elastic with no activation, where Young's modulus (E_L) was derived from the longitudinal elastic shear modulus (μ_L) measured at 0-5% gripping force (contact without tightening condition) and the longitudinal Poisson's ratio (υ_L) was derived from [7]; (ii) isotropic elastic with maximum activation, conducted in the same way as the previous case but at 30% gripping force; and (iii) transversally isotropic elastic with maximum activation, where the muscle behaviour was considered to be anisotropic, with rotational symmetry around the fibre axis. A shear modulus, Young's modulus, and Poisson's ratio were required for the longitudinal (μ_L, E_L, υ_L) and transverse (μ_T, E_T, υ_T) directions [7]. All the previous parameters stemmed from the shear elastic moduli measured at 30% gripping force, and the other parameters were deduced from [7].

3. Results

3.1. Effects of Gripping Force on Muscle Elasticity and Fibre Anisotropy

Elastography measurements highlighted that the shear elastic modulus of the FDIM evolved differently depending on the fibre orientation (Figure 2a,b). Parallel to the fibres, the shear modulus increased by more than 12 times between 0 and 30% strength, with a strong increase between 5 and 15% (Figure 2c). Perpendicular to the fibres, the modulus remained almost constant with the gripping force (Figure 2c). These measurements allowed us to identify the parameters of the FE model described in Table 1.

Table 1. Muscle properties used in the FE simulations.

	Isotropic Muscle		Anisotropic Muscle with Maximum Activation	
	No Activation	**Maximum Activation**	**Longitudinal**	**Transversal**
Shear elastic modulus	Not used	Not used	$\mu_L = 121.0$ kPa	$\mu_T = 9.0$ kPa
Young's modulus	$E_L = 22.4$ kPa	$E_L = 338.8$ kPa	$E_L = 338.8$ kPa	$E_T = 25.2$ kPa
Poisson's ratio	$\upsilon_L = 0.499$	$\upsilon_L = 0.499$	$\upsilon_L = 0.499$	$\upsilon_T = 0.963$

3.2. Effects of Muscle Activation on DPMI and Local Transmissibility

The resonance of the wrist around 35 Hz is observable in both the model and standard ISO 10068 DPMI. At higher frequencies, the model differs from the standard (Figure 3a). Muscle activation had a marked effect on the global DPMI and local transmissibility beyond 100 Hz, changing the amplitude and frequency of the peaks as well as the spatial nature of the harmonic response in the muscle (e.g., Figure 3b maps).

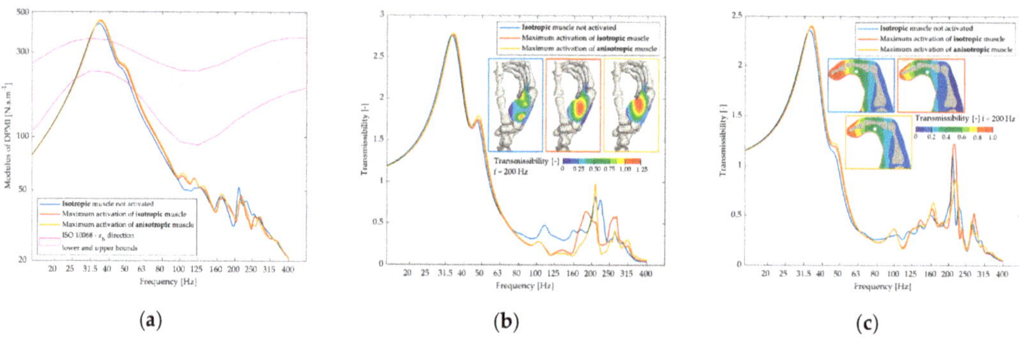

Figure 3. (**a**) DPMI of FE model compared to standard ISO 10068. (**b**) Averaged vibration transmissibility between the handle and the first dorsal interosseous muscle. (**c**) Vibration transmissibility between the handle and the tissue under the median phalanx of the index (white dot).

4. Discussion and Conclusions

The SSI technique demonstrated an ability to measure the mechanical elasticity of the FDIM. It revealed that the evolution of stiffness is strongly dependant on muscle activation and fibre orientation. Many sources of uncertainty arose during the measurements. For example, muscle heterogeneity (Figure 2a) and probe orientation, with regard to fibre direction, may be responsible for measurement discrepancies. Particular attention should be given to rendering the apparatus more robust before extending the measures to a cohort of subjects.

The model showed that muscle activation significantly affects the transmission of vibrations beyond 100 Hz. The gap between the DPMI computed by our model and that of ISO 10068 may be explained by the numerous simplifications made. More realistic boundary conditions should be applied to the wrist. In addition, modal superposition may only take into account basic viscoelastic effects. Hence, more complex rheological models will have to be used with direct resolution methods or by using our model in the time domain.

In conclusion, we succeeded in using the SSI technique to link active hand muscle elasticity with gripping force to feed a complex FE hand model. Our model allowed us to compute dynamic responses to vibrations and quantify the effect of local muscle activation on vibration propagation within the hand.

Author Contributions: Conceptualisation, S.V., C.N. and E.J.; methodology, S.V., C.N., H.H.P.N., J.-L.G., J.C., E.F., E.J. and N.S.; software, S.V.; validation, S.V., C.N., H.H.P.N. and E.J.; formal analysis, S.V., C.N. and E.J.; investigation, S.V., C.N., E.J. and N.S.; writing—original draft preparation, S.V.; writing—review and editing, C.N., E.J., N.S., H.H.P.N., J.-L.G., J.C. and E.F.; visualisation, S.V.; supervision, C.N. and E.J. All authors have read and agreed to the published version of the manuscript.

Funding: This research received no external funding.

Institutional Review Board Statement: The study was conducted according to the guidelines of the Declaration of Helsinki and approved by the Institutional Review Board of the FRENCH NATIONAL AGENCY FOR MEDICINES AND HEALTH PRODUCTS (No. ID-RCB 2022-A01616-37, 07/11/22) and by the FRENCH NATIONAL ETHICAL RESEARCH COMMITTEE (CPP 22.03423.000122, 07/11/22).

Informed Consent Statement: Informed consent was obtained from all the subjects involved in the study.

Data Availability Statement: The data that support the findings of this study are available from the corresponding author upon reasonable request.

Conflicts of Interest: The authors declare no conflict of interest.

References

1. *FD CR 12349:1996*; Mechanical Vibration-Guide to the Health Effects of Vibration on the Human Body. AFNOR: Paris, France, 1996.
2. *NF EN ISO 5349-1:2001*; Mechanical Vibration-Measurement and Evaluation of Human Exposure to Hand-Transmitted Vibration-Part 1: General Requirements. AFNOR: Paris, France, 2002.
3. Aldien, Y.; Marcotte, P.; Rakheja, S.; Boileau, P.E. Influence of Hand-Arm Posture on Biodynamic Response of the Human Hand-Arm Exposed to Zh-Axis Vibration. *Int. J. Ind. Ergon.* **2006**, *36*, 45–59. [CrossRef]
4. Bouillard, K.; Nordez, A.; Hug, F. Estimation of Individual Muscle Force Using Elastography. *PLoS ONE* **2011**, *6*, e0029261. [CrossRef] [PubMed]
5. Noël, C.; Settembre, N.; Reda, M.; Jacquet, E. A Multiscale Approach for Predicting Certain Effects of Hand-Transmitted Vibration on Finger Arteries. *Vibration* **2022**, *5*, 213–237. [CrossRef]
6. *ISO 10068:2012*; Mechanical Vibration and Shock-Mechanical Impedance of the Human Hand-Arm System at the Driving Point. ISO: Geneva, Switzerland, 2012.
7. Rouze, N.C.; Wang, M.H.; Palmeri, M.L.; Nightingale, K.R. Finite Element Modeling of Impulsive Excitation and Shear Wave Propagation in an Incompressible, Transversely Isotropic Medium. *J. Biomech.* **2013**, *46*, 2761–2768. [CrossRef] [PubMed]

 *proceedings*

Proceeding Paper

Vibration Emissions of Grinders: Experiments and a Model †

Quentin Pierron 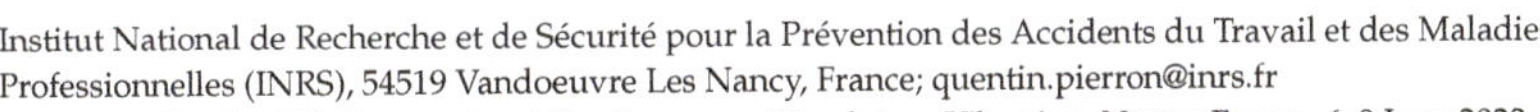

Institut National de Recherche et de Sécurité pour la Prévention des Accidents du Travail et des Maladies Professionnelles (INRS), 54519 Vandoeuvre Les Nancy, France; quentin.pierron@inrs.fr

† Presented at the 15th International Conference on Hand-Arm Vibration, Nancy, France, 6–9 June 2023.

Abstract: A simple two-rigid-body model of an electrical grinder was created to calculate the vibration emissions according to the test code EN 60745-2-3. The model assumes that the operator does not affect the vibration emissions. Experiments with and without the operator validated this hypothesis and demonstrated the model's abilities and limitations.

Keywords: hand-arm vibration; grinder; model; test code

Citation: Pierron, Q. Vibration Emissions of Grinders: Experiments and a Model. *Proceedings* **2023**, *86*, 13. https://doi.org/10.3390/proceedings2023086013

Academic Editors: Christophe Noël and Jacques Chatillon

Published: 10 April 2023

1. Introduction

Electric grinders are vibrating tools that are likely to expose operators to vibration levels higher than the values defined by the European directive. To help companies to choose machines with fewer vibrations, manufacturers are required to declare the vibration emissions. The vibration emissions constitute the frequency-weighted root-mean-square acceleration a_{hv} defined by the standard ISO 5349-1 [1] (typically between 3 and 8 m/s^2 [2–6] for grinders), measured in accordance with the standard test code EN 60745-2-3. The objective of this study was to develop and verify a simple model which calculates the accelerations of a grinder during standard tests. This model assumes that holding the grinder has no effect on vibration [7,8]. To validate this hypothesis, in addition to standard tests during which an operator holds the grinder, tests without the operator were carried out. Finally, the measured accelerations were compared to those calculated by the model.

2. Materials and Methods

According to the standard EN 60745-2-3 [9], the disks were created from aluminum and perforated in such a way as to create the given imbalance. A thin steel cable was passed through a pulley and attached to the grinder and a mass so as to apply an upwards force and to relieve the weight of the grinder. The grinder was held by an operator and when the grinder was running, the disk spun freely without grinding or cutting any material. Three operators handled the grinders successively and repeated the tests. Other tests were carried out without an operator, and the grinder was simply hung from a long and rigid 1930 N/m tension spring.

The tested grinder was a Metabo W12-125 Quick (Nürtingen, Germany). This is a small grinder weighing 2.4 kg with 125 mm diameter disks. The grinder was tested without the side handle.

For the tests, the accelerations were measured using at least three PCB 356B21 glued triaxle piezoelectric accelerometers. The tension signals of the accelerometers were numerically recorded using a Dewesoft R2DB (Trbovlje, Slovenia) at a frequency of 20,000 Hz. Before the tests, the respective positions of the accelerometers were measured through a 3D scan of the instrumented grinders.

3. Numerical Model

A numerical model was developed considering the perforated disk and the grinder as two rigid bodies linked by a revolute joint. A constant rotational speed ω between them was assumed.

The main hypothesis of this model was the lack of external force acting on the grinders. Finally, the dynamic equations provided a system of six equations and six degrees of freedom (position of the grinder in the space, including the vector of angular velocity, denoted as $\vec{\Omega}$). The system was simplified by neglecting the second-order terms in the angular velocity ($\sim \Omega^2$) and other terms through the comparison of the numerical values of mass parameters. Lastly, the complex amplitude vector of the first harmonic (at the rotation frequency) of the steady-state acceleration of point M on the running grinder was given by the two following equations:

$$\vec{a}_M = \frac{\text{bal}}{m_T}\omega^2\vec{e}_r + j\omega\vec{\Omega} \wedge \left(\vec{CM} - \frac{m_G}{m_T}\vec{CG}_{woD} \right) \tag{1}$$

$$\overline{\overline{I}}_T\vec{\Omega} = j\omega m_G \frac{\text{bal}}{m_T}\vec{CG}_{woD} \wedge \vec{e}_r \tag{2}$$

In these equations, bal is the imbalance of the perforated disk, m_T is the total mass of the system, $\vec{e}_r$ is the rotating vector in the plane of the disk, $j = \sqrt{-1}$, C is the center of the disk, m_G is the mass of the grinder without the disk, G_{woD} is the center of mass of the grinder without the disk, and $\overline{\overline{I}}_T$ is the sum of the inertia tensor of rigid bodies expressed at their center of mass. All model parameters were known and provided by the manufacturer based on their detailed CAD.

4. Results

Firstly, the frequency-weighted root-mean-square accelerations a_{hv} defined in the standard ISO 5349-1 and implemented during the use of the test codes and tests with the hanging grinder are shown in Figure 1. The accelerations a_{hv} were very close, regardless of whether the grinder was hanging or held by an operator. The low standard deviation values for the test code confirm that handling had no effect on the vibrations. One of the hypotheses regarding the model was therefore valid for this grinder.

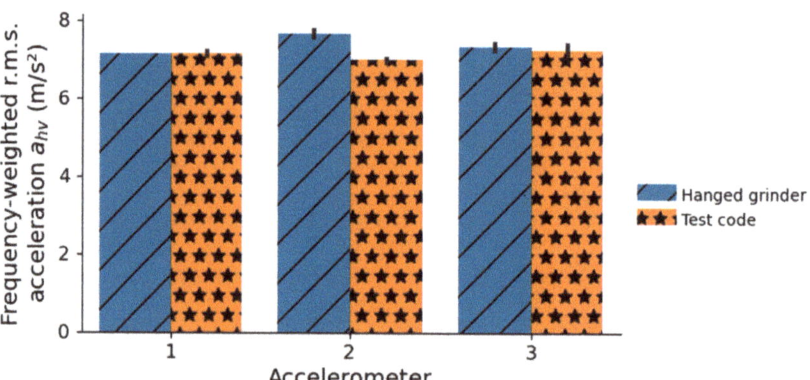

Figure 1. Frequency-weighted root-mean-square accelerations a_{hv} measured during the use of the test codes and tests with the hanging grinder. The vertical black lines indicate the standard deviation.

Secondly, the amplitude vector of the first harmonic regarding the acceleration was calculated based on the location of the accelerometers. This acceleration was compared to the experimental acceleration obtained with the hanging grinder, as shown in Figure 2.

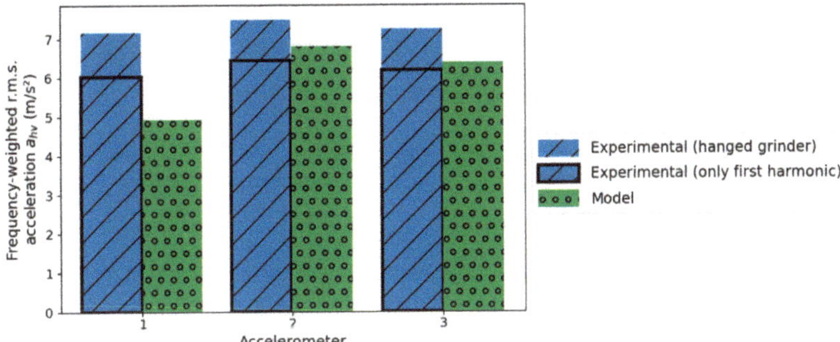

Figure 2. Frequency-weighted root-mean-square acceleration a_{hv}: measurement with three accelerometers taking into consideration the entire signal or only the first harmonic and numerical results of the model at the locations of the accelerometers.

5. Discussion

The tests described above were conducted using a grinder without a flexible part. In addition, similar tests were carried out using a Bosch GWS 24-230 LVI grinder (Gerlingen, Germany) with both handles being flexible (vibration reduction system). This second grinder was large, weighing 5.5 kg with 230 mm diameter disks. The comparisons of the measured frequency-weighted root-mean-square accelerations a_{hv} in the two tests with (test code) and without an operator (hanging grinder) are shown in Figure 3. The vibrations were not linked to whether an operator was holding the grinder or not, except for the rear handle, which is flexible. The results show that a flexible rear handle reduces vibration exposure, as expected. The model did not account for the flexibility of the grinder parts, and therefore, it cannot be used to calculate the correct acceleration on the flexible handle. However, it could replace the test without an operator to evaluate the effect of the flexible part and could aid in the design of the flexible part.

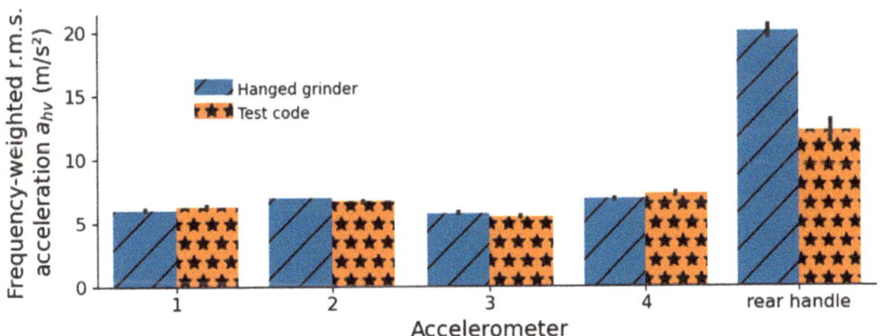

Figure 3. Frequency-weighted root-mean-square accelerations a_{hv} measured during the use of the test codes and tests with the hanging grinder for the second grinder. The vertical black lines indicate the standard deviation.

6. Conclusions

Comparing the vibration a_{hv} on a hanging grinder to that obtained according to the test code EN 60745-2-3, no difference was observed. Thus, for this type of grinder, there is no need to consider the operator's hand in order to model and predict the vibration emissions of grinders. A simple two-rigid-body model was developed and allowed us to calculate this vibration value, as declared by the manufacturer. This model cannot be applied to grinders containing flexible parts.

Funding: This research received no external funding.

Institutional Review Board Statement: Not applicable.

Informed Consent Statement: Not applicable.

Data Availability Statement: Not applicable.

Conflicts of Interest: The author declares no conflict of interest.

References

1. *ISO 5349-1:2001*; Mechanical Vibration—Measurement and Evaluation of Human Exposure to Hand-Transmitted Vibration—Part 1: General Requirements. International Organization for Standardization: Geneva, Switzerland, 2001.
2. Mirbod, S.M.; Inaba, R.; Iwata, H. A Study on the Vibration-Dose Limit for Japanese Workers Exposed to Hand-Arm Vibration. *Ind. Health* **1992**, *30*, 1–22. [CrossRef] [PubMed]
3. Burström, L.; Lundström, R.; Hagberg, M.; Nilsson, T. Comparison of Different Measures for Hand–Arm Vibration Exposure. *Saf. Sci.* **1998**, *28*, 3–14. [CrossRef]
4. Jang, J.-Y.; Kim, S.; Park, S.K.; Roh, J.; Lee, T.-Y.; Youn, J.T. Quantitative Exposure Assessment for Shipyard Workers Exposed to Hand-Transmitted Vibration From a Variety of Vibration Tools. *AIHA J.* **2002**, *63*, 305–310. [CrossRef] [PubMed]
5. Rimell, A.N.; Notini, L.; Mansfield, N.J.; Edwards, D.J. Variation between Manufacturers' Declared Vibration Emission Values and Those Measured under Simulated Workplace Conditions for a Range of Hand-Held Power Tools Typically Found in the Construction Industry. *Int. J. Ind. Ergon.* **2008**, *38*, 661–675. [CrossRef]
6. Edwards, D.J.; Rillie, I.; Chileshe, N.; Lai, J.; Hosseini, M.R.; Thwala, W.D. A Field Survey of Hand–Arm Vibration Exposure in the UK Utilities Sector. *Eng. Constr. Archit. Manag.* **2020**, *27*, 2179–2198. [CrossRef]
7. Lemerle, P.; Klingler, A.; Trompette, N.; Cristalli, A.; Geuder, M. Development and Validation of an Accurate Testing Procedure to Measure Coupling Forces and Characterize the Man/Machine Interaction. In Proceedings of the 11th International Conference on Hand-Arm Vibration, Bologna, Italy, 3–7 June 2007; pp. 3–7.
8. Liljelind, I.; Pettersson, H.; Nilsson, L.; Wahlström, J.; Toomingas, A.; Lundström, R.; Burström, L. Determinants Explaining the Variability of Hand-Transmitted Vibration Emissions from Two Different Work Tasks: Grinding and Cutting Using Angle Grinders. *Ann. Occup. Hyg.* **2013**, *57*, 1065–1077. [PubMed]
9. *EN 60745-2-3:2011*; Hand-Held Motor-Operated Electric Tools—Safety—Part 2-3: Particular Requirements for Grinders, Polishers and Disk-Type Sanders. CENELEC: Brussels, Belgium, 2011.

Proceeding Paper

Interference of Vibration Exposure in the Force Production of the Hand–Arm System †

Massimo Cavacece [1,*] , **Angelo Tirabasso** [2], **Raoul Di Giovanni** [2], **Stefano Monti** [2], **Enrico Marchetti** [2,*] and **Luigi Fattorini** [3]

1   Department of Civil and Mechanical Engineering, University of Cassino and Southern Lazio, 03043 Cassino, Italy
2   INAIL, DIMEILA, Laboratory Physical Agents, 00078 Monte Porzio Catone, Italy; a.tirabasso@inail.it (A.T.); r.digiovanni@inail.it (R.D.G.); st.monti@inail.it (S.M.)
3   Department of Physiology and Pharmacology « V. Erspamer », Sapienza University of Roma, 00185 Rome, Italy; luigi.fattorini@uniroma1.it
*   Correspondence: cavacece@unicas.it (M.C.); e.marchetti@inail.it (E.M.)
†   Presented at the 15th International Conference on Hand-Arm Vibration, Nancy, France, 6–9 June 2023.

Abstract: The authors evaluated the short-term neuromuscular effects on the assessment of mechanical hand—arm systems induced by vibrating tools to investigate the relationship between the force exerted and the vibration exposure. The motor task consisted of holding the instrumented handle with the dominant hand at predetermined grip force values. Five subjects took part in the tests. The tests were developed in the absence of vibration and in the presence of vibration at 5 m/s^2, 7.5 m/s^2 and 10 m/s^2. The push and pull force values were calculated in the tests on the five subjects.

Keywords: tonic vibration reflex; mechanical vibrations; exhaustive motor task; push–pull force control

Citation: Cavacece, M.; Tirabasso, A.; Di Giovanni, R.; Monti, S.; Marchetti, E., Fattorini, L. Interference of Vibration Exposure in the Force Production of the Hand–Arm System. *Proceedings* **2023**, *86*, 36. https://doi.org/10.3390/proceedings2023086036

Academic Editors: Christophe Noël and Jacques Chatillon

Published: 20 April 2023

1. Introduction

The active behavior of the tonic vibration reflex (TVR) in the assessment of vibration exposure in the mechanical hand–arm system (HAS) is well-known [1–3]. The interaction between vibration and muscular contraction may generate disturbance in motor tasks and force development. This paper aimed to assess the force parameters during exhaustive grip force tasks with and without vibration. The hypothesis is that vibration exposure can induce early fatigue and unbalanced motor control during a motor task.

2. Material and Methods

The authors evaluated the short-term neuromuscular effects on the HAS induced by vibrating tools to investigate the relationship between forces exerted and vibration exposure [1]. The tests were performed at 30 Hz at different accelerations with a grip force of 30% of the maximum voluntary contraction (MVC), as explained below. The selected frequency represents the frequency inducing maximal hand–arm energy transmission, as reported in Fattorini et al. [2,3].

2.1. Subjects

Five subjects took part in the tests (Table 1). Each volunteer underwent the MVC measurement, exerting grip force with the dominant hand on the handle of the shaker, switched off, three times. The average of the measurements offered the MVC value of each subject. The grip force values depend on the average value [2].

Table 1. Characteristics of Subjects and MVC values.

Subject	Subject				
	A	**B**	**C**	**D**	**E**
Height [cm]	170	167	170	184	181
Weight [kg]	103	72	65	93	88
Gender	M	M	M	M	M
MVC 1 [N]	470	400	380	470	330
MVC 2 [N]	490	430	360	460	310
MVC 3 [N]	490	400	350	450	330
MVC 30% [N]	140	120	110	140	100

2.2. Motor Task

The motor task consisted of holding the instrumented handle with the dominant hand at predetermined grip force values. The handle had two strain gauges, measuring push and pull forces, and the subject had to maintain the target force value for as long as possible (Table 1). To measure both components of gripping force (i.e., push and pull) the handle was divided into two halves, as described in Fattorini et al. [2]. The deformation of the handle resulted in a strain gauge response. This configuration allowed for continuous control of push and pull forces on an oscilloscope positioned in front of the operator. Temperature and humidity were maintained by an air conditioner to stabilize the strain gauges' transfer function. Before the tests, MVC was evaluated as the maximal force between three trials of maximal gripping. During this evaluation, the subject's posture was that described in Fattorini et al. [2]. The subject stood on an elevated platform to adjust the forearm and handle axes. The subject was instructed to balance push and pull force to attain pure grip force, without any possible component from the shoulder. Both components of grip force were continuously recorded and displayed to the subject by an oscilloscope (Hewlett & Packard, 54603B, Palo Alto, CA, USA) to help maintain a fixed level of force and balance between push and pull. The test consisted of exerting grip force at the level defined in Table 1 for as long as possible, with and without vibration. Different percentages of MVC tests on the same subject were randomized to avoid hysteresis (Table 1). Performing the grip action on the instrumented handle can induce HAS fatigue, which can yield a set of changes in the HAS's regulation systems. Therefore, the minimum rest period between successive tests was 60 min. The rest period could increase with the value of the intensity of the gripping force exerted on the handle [4].

3. Results

The coordinate system adopted was the biodynamic one of UNI EN ISO 5349-1 [5]. The experiment was along the direction of the z axis. The coordinates were measured concerning the human hand while holding a cylindrical handle. The acquisition system eliminated sample aliasing effects during the digitization of the acquired signals using an analog-to-digital converter (ADC). Push and pull forces represent a complex mechanism in the presence of mechanical vibrations on the HAS. The experimental investigations considered three sinusoidal signals at 30 Hz with an r.m.s. acceleration of 5 m/s^2, 7.5 m/s^2 and 10 m/s^2. Subjects developed a pull and push strength level of 30% of their *MVC* value. The difference between the pull and push force was assessed by the following relationship:

$$\Delta G = \left[\frac{(Mean\ Push\ Force - Mean\ Pull\ Force)}{MVC\ 30\%} \right] \tag{1}$$

The difference ΔG between the pull and push force is lower in the test with a vibration level of 5 m/s^2 than in tests with a vibration level of 7.5 m/s^2 or 10 m/s^2 (Figure 1). The ΔG value and the standard deviation assume lower values with the vibration level of

5 m/s^2 than with the vibration levels of 7.5 m/s^2 and 10 m/s^2 (Figure 2). The handle endurance times on the shaker are shown in Table 2.

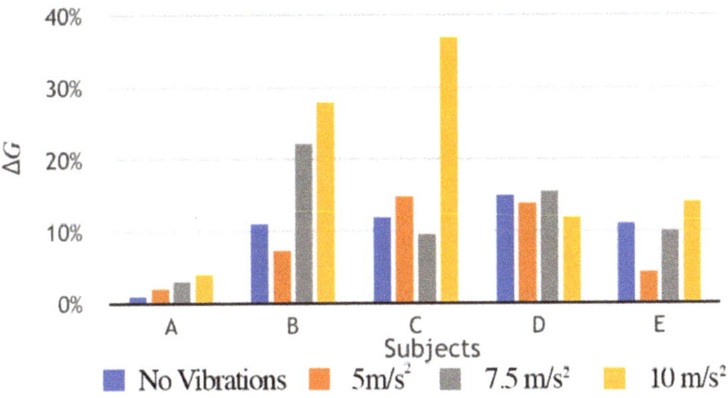

Figure 1. Values of ΔG evaluated on 5 subjects in the absence of vibration and in the presence of vibration with accelerations at 5 m/s^2, 7.5 m/s^2 and 10 m/s^2.

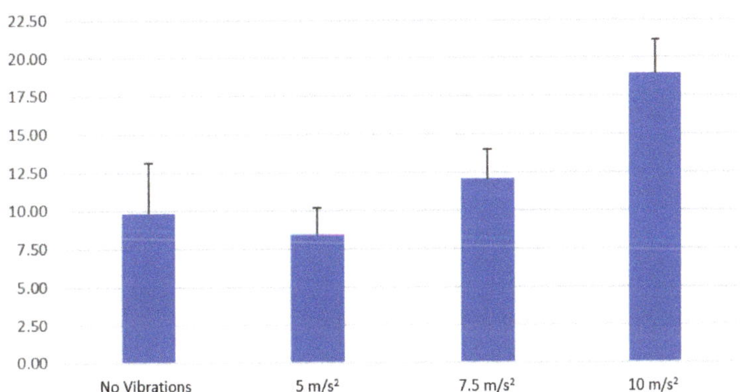

Figure 2. The ΔG value and the standard deviation evaluated in the absence and presence of vibration with accelerations of 5 m/s^2, 7.5 m/s^2 and 10 m/s^2.

Table 2. Time of gripping maintenance without and with vibration.

Subject	No	Vibrations		
	Vibrations	$5.0 \text{ [m/s}^2]$	$7.5 \text{ [m/s}^2]$	$10.0 \text{ [m/s}^2]$
		Time [s]		
A	205	146	140	145
B	264	312	242	213
C	288	204	202	280
D	275	307	286	307
E	303	305	302	295

4. Discussion

In humans, force production is a complex task involving the nervous and muscular systems. The former sends a command, i.e., the motor drive, to the muscular apparatus, which responds with the contraction. In the presence of any external factor, the proprioceptive apparatus, input via the nervous system, reveals this occurrence, and the motor drive is rearranged. This paper aimed to evaluate the influence of the vibration, as an external factor, on force parameters, fatigue and the push–pull balance in a controlled grip task. The present fatigue results did not show evidence of changes in the time of force exertion with vibration. Indeed, as reported in Table 2, the time duration before fatigue is quite unchanged with vibration compared to without vibration. These findings are likely due to the neuromuscular system's capacity to modulate muscular contractions to always perform the gripping task, even in different working conditions. In this regard, it must be considered that the gripping task involves a great number of muscles belonging to different anatomical districts, such as the hand, forearm, arm and shoulder, other than the anti-gravitational ones. It is conceivable that the nervous system can modulate the force of every muscle involved to obtain always the target output force, and while maintenance endurance time is unchanged, the muscle interplay is probably changed. The change in muscle interplay could be observed by measuring the push and pull force in the gripping task. In Figure 2, these measurements in the different experimental conditions are shown. It is quite evident from the figure that the vibration must be considered a sort of noise for the nervous system. The nervous system can perform the target task in all conditions but with different muscular interplay engagement. In particular, it is possible to affirm a different behavior, an unbalance, of the forearm muscles being responsible for the production of push and pull forces. Moreover, the unbalance is related to the vibration's acceleration. This finding was expected because more acceleration corresponds to greater input and, in turn, more noise.

Finally, our findings show a lower level of coordination between the two components of grip force: push and pull.

5. Conclusions

Force production parameters, fatigue and push and pull force values were assessed with and without vibration on five subjects. Vibration does not seem to influence the fatigue phenomenon because of a neuromuscular rearrangement. These changes were recognized by ΔG values representing the push and pull balance during the gripping task. These findings show a clear relationship with vibration acceleration. Present data confirm the neuromuscular plasticity involved in adapting the force production in interfering conditions at the dispense of fine muscle control. The loss of fine muscle control should be better investigated to monitor muscular integrity.

Author Contributions: Conceptualization, M.C., E.M. and L.F.; methodology, L.F. and A.T.; software, R.D.G.; validation, E.M., L.F. and M.C.; formal analysis, R.D.G. and S.M.; resources, E.M. and A.T.; data curation, S.M. and R.D.G.; writing—original draft preparation, M.C.; writing—review and editing, E.M. and L.F.; funding acquisition, E.M. All authors have read and agreed to the published version of the manuscript.

Funding: This research received no external funding.

Institutional Review Board Statement: Not applicable.

Informed Consent Statement: Informed consent was obtained from all subjects involved in the study.

Data Availability Statement: Not applicable.

Conflicts of Interest: The authors declare no conflict of interest.

References

1. Martin, B.J.; Parc, H.S. Analysis of the tonic vibration reflex: Influence of vibration variables on motor unit synchronization and fatigue. *Eur. J. Appl. Physiol.* **1997**, *75*, 504–511. [CrossRef] [PubMed]
2. Fattorini, L.; Tirabasso, A.; Lunghi, A.; Di Giovanni, R.; Sacco, F.; Marchetti, E. Muscular forearm activation in hand-grip tasks with superimposition of mechanical vibrations. *J. Electromyogr. Kinesiol.* **2016**, *26*, 143–148. [CrossRef] [PubMed]

3. Fattorini, L.; Tirabasso, A.; Lunghi, A.; Di Giovanni, R.; Sacco, F.; Marchetti, E. Muscular synchronization and hand-arm fatigue. *Int. J. Ind. Ergon* **2017**, *62*, 13–16. [CrossRef]
4. Cavacece, M. Incidence of Predisposing Factors on the Human Hand-arm Response with Flexed and Extended Elbow Positions of Workers Subject to Different Sources of Vibrations. *Univers. J. Public Health* **2021**, *9*, 507–519. [CrossRef]
5. *ISO 5349-1:2004*; Mechanical Vibration Measurement and Evaluation of Human Exposure to Hand-Transmitted Vibration, Part 1: General Requirements. International Organization for Standardization: Geneva, Switzerland, 2004.

proceedings 

Proceeding Paper

Using an Impact Wrench in Different Postures—An Analysis of Awkward Hand–Arm Posture and Vibration [†]

Nastaran Raffler *[ID], Thomas Wilzopolski and Christian Freitag

Institute for Occupational Safety and Health of the German Social Accident Insurance (IFA), Alte Heerstr. 111, 53757 Sankt Augustin, Germany
* Correspondence: nastaran.raffler@dguv.de; Tel.: +49-30-13001-3432
† Presented at the 15th International Conference on Hand-Arm Vibration, Nancy, France, 6–9 June 2023.

Abstract: Overhead work and awkward hand–arm posture can impact muscle load. Additional workloads such as hand–arm vibration exposure while carrying or holding a power tool can contribute to adverse health effects. This study investigated the posture and muscle activity of 11 subjects while using an impact wrench in three working directions: upwards, forwards, and downwards. Although the vibration exposure did not show notable differences in the magnitude (4.6 ± 0.2 m/s^2), postural behaviour and muscle activity showed higher workloads for working upwards and downwards compared to forwards. The results of muscle activity, and the self-reported exposure level, highlight the necessity of considering posture while exposed to vibration exposure.

Keywords: awkward posture; hand–arm vibration; electromyography

1. Introduction

From an ergonomic perspective, overhead work should be avoided. When an awkward body posture is adopted, stress is placed on the musculoskeletal system, which can cause irreversible damage to a person's health. Additional exposure to hand–arm vibrations (HAV) can increase the musculoskeletal system's workload and result in adverse health effects. The extent to which body posture affects the impact of vibrations on the hand–arm system has not yet been studied in detail. However, many recent studies highlight the effect of combined exposures to posture and vibration.

Tayler et al. [1] showed that the measured vibration magnitude on a tool handle neither relates to the perception of the exposure nor to the hand-transmitted vibration in varying postures across the test participants. Additionally, other studies [2,3] have shown that flexion of the elbow and wrist position seem to affect the dissipation of HAV exposure and the power absorption into the hand–arm system.

Furthermore, electromyography (EMG) technique is an additional tool used to investigate the surface electric muscle activity. This allows for a detailed analysis of the combined workloads of awkward posture and vibration on the muscles. Therefore, in this study, three different working directions have been investigated in terms of the hand–arm vibration magnitude, shoulder and hand–arm posture, and the response of the body in the form of electric muscle activity.

2. Materials and Methods

2.1. Subjects and Experimental Procedure

A total of 11 healthy, right-handed volunteers (4 female and 7 male, average $\pm$ standard deviation = 36 ± 11 years old; 178 ± 8 cm high and 76 ± 15 kg) participated in this project.

Three directions (downwards, forwards, and upwards) were chosen in order to investigate the influence of the working direction while using an impact wrench (Figure 1). A height-adjustable test setup was designed to enable equal starting conditions for each test

Citation: Raffler, N.; Wilzopolski, T.; Freitag, C. Using an Impact Wrench in Different Postures—An Analysis of Awkward Hand–Arm Posture and Vibration. *Proceedings* **2023**, *86*, 40. https://doi.org/10.3390/proceedings2023086040

Academic Editors: Christophe Noël and Jacques Chatillon

Published: 26 April 2023

person (90-degree forearm angle perpendicular to the 100 mm × 800 mm × 300 mm oak plank). 100-mm-long wood screws were screwed in 12 times for each direction. The power tool was the Bosch Professional GDX 18V-LI impact wrench.

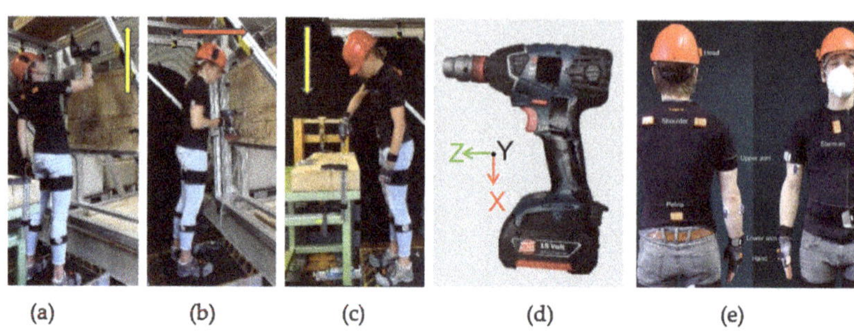

Figure 1. (**a–c**): Test setup for examining the three working directions, indicated by arrows, (**d**): impact wrench with a triaxial accelerometer, (**e**): wireless measurement system for body posture (sensor attachment).

2.2. Hand–Arm Vibration and Subjective Perception of the Exposures

In accordance with ISO 5349-1 and 2, the tool handle vibration was measured for a period of 12 screwdriving operations. Accelerometers were glued on the tool handle in accordance with ISO 28927-5 (Figure 1d). The vibration magnitudes were expressed as root-mean-square acceleration, which was frequency-weighted and band-limited using the W_h filter. The vibration total value a_{hv} of the frequency-weighted acceleration values for the x, y, and z axes are calculated using the following:

$$a_{hv} = \sqrt{a_{hwx}^2 + a_{hwy}^2 + a_{hwz}^2} \tag{1}$$

To analyse the perceived vibration and posture exposure for the individual test sections, the Borg CR10 scale was used: 0 for absolutely nothing to 10 for extremely strong.

2.3. Posture

To measure body posture data, the Xsens Awinda wireless measurement system (18 inertial measurement units) with 60 Hz measurement frequency was used (Figure 1e). After the recording of the body posture (MVN 2022.0 Analyse), the data was imported into WIDAAN, an analysing software developed by The Institute for Occupational Safety and Health of the German Social Accident Insurance [4]. In WIDAAN, the angle of the body is analysed in accordance with DIN EN 1005-4. The amount of time a body angle is within a range of motion is specified as a percentage of the total measurement. The following angles (Table 1) will be presented in this paper, along with the associated categories for neutral, moderate, and awkward movement ranges:

Table 1. Investigated body angles with associated categories for movement ranges.

Category	Shoulder Flexion	Upper-Arm Inclination	Wrist Flexion	Wrist Radialduktion
neutral	0–20°	0–20°	−25–20°	−10–10°
moderate	20–60°	20–60°	−25––50° or 20–45°	−10––25° or 10°–15°
awkward	<0° or >60°	<0° or >60°	<−50° or >45°	<−25° or >15°

2.4. Electromyography

A wireless surface electromyography measuring system, Cometa Wave Plus (sampling rate 2000 Hz) has been used to analyse the electrical muscle activity. Four transducers were placed on the right-hand side of the hand–arm system. Following Hansson et al.1997 [5], a bandpass filtering of 30–400 Hz and a RMS calculation was carried out with a rectangular window (0.125 s). A percentage value is calculated based on the maximum voluntary contraction (MVCP values) of the individual muscles. The electrodes were positioned at the following muscles: musculus trapezius descendens, musculus biceps brachii, musculus flexor capri ulnaris, and musculus extensor digitorum.

2.5. Data Analysis and Statistics

For the evaluations, only the screwing-in processes were included with no pauses. The EMG and HAV measurement data were imported to the body posture analysis software "WIDAAN", followed by synchronisation with body angles and video data.

3. Results

3.1. Experimental Procedure

On average, 6.3 s were required for the forwards direction, while the upwards and downwards directions required 6.0 and 5.8 s, respectively.

3.2. Hand–Arm Vibration and Subjective Perception of the Exposures

The vibration total values for the upwards (4.8 m/s^2) and downwards (4.7 m/s^2) directions are comparable. For the forwards working direction, a total vibration value of 4.4 m/s^2 was recorded (Figure 2a).

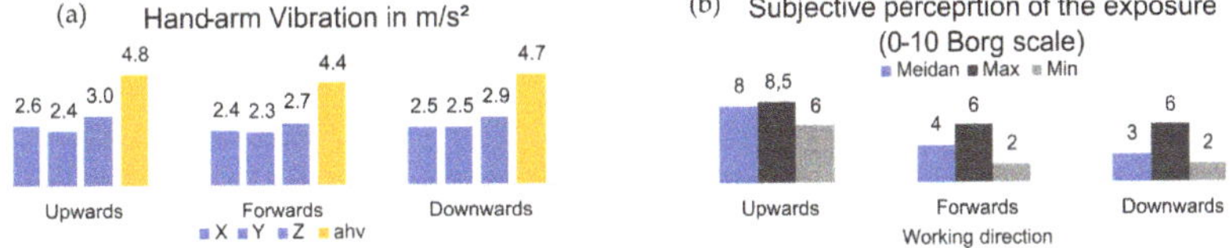

Figure 2. (**a**) Vibration exposure for x, y, and z axis and the vibration total value; (**b**) subjective perception for the exposure of vibration and posture in three working directions.

Figure 2b shows the test subjects' perceived rating of the vibration and posture exposure, using the Borg scale. While the test subjects rated the combined exposures as "very strong", when working in the upwards direction (median 8), the combined exposures in the forwards (median 4) and downwards direction (median 3) were rated as medium.

3.3. Posture

Figure 3a–d shows the recorded body angles while using the impact wrench as box plots (5th to 95th percentile). Concerning shoulder flexion, upper arm inclination, and wrist flexion, working in the upwards direction shows more percentages in the awkward and moderate range of movements than the other directions. For wrist radialduction (Figure 3d), the data suggest a neutral-to-moderate range when working in the upwards and forwards working directions. However, when working in the downwards direction, the data for radialduction was very different to the other working directions. In this direction, radialduction was entirely in the moderate risk category, whereas the direction of radialduction changed completely to ulnarduction (outwards, negative values).

3.4. Electromyography

In Figure 3e–h, the EMG data are grouped according to muscles and working directions. The highest MVCP values were observed by working upwards. However, for the extensor digitorum (Figure 3h), the highest level of muscle activity was in the downwards working direction (median 74%).

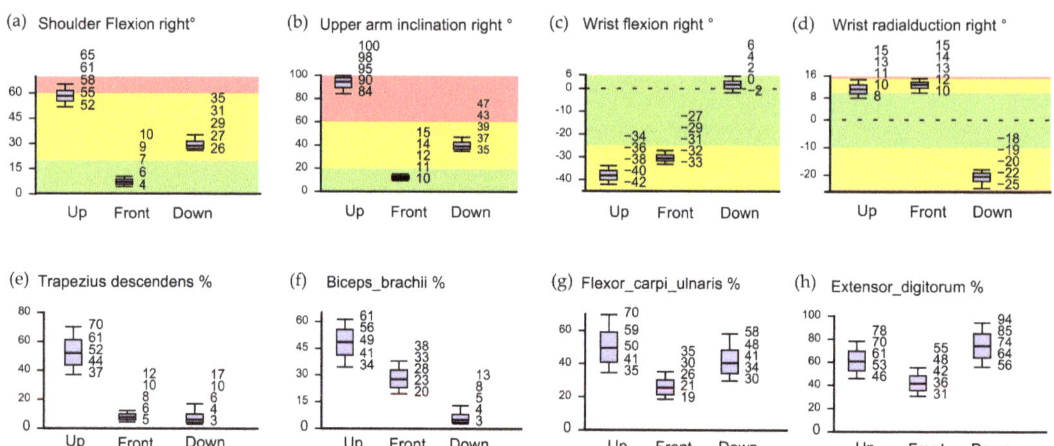

Figure 3. (**a–d**): Posture analysis for three working directions while using an impact wrench for different angles, colour coding represents: green for neutral posture, yellow for moderate posture and red for awkward posture, (**e–h**): electro muscle activity in three working directions for different muscles.

4. Discussion

The overall vibration values from the power tool are just under the limit of 5 m/s^2 [EU-Directive 2002/44/EC 2002] and did not show a noticeable difference in magnitude in the three different directions. When looking at the risk categories, it is evident that the test subjects frequently adopted an awkward posture when working in the upwards direction. When working in the downwards direction, the level of wrist flexion is improved, while a high level of ulnarduction is seen since the wrist needs to be in a highly angled position for the downwards screwdriving motion. These findings line up with those of the studies carried out by Besa et al. [3] in that they also highlight the impact of hand–arm posture on the transmission of energy to the hand–arm system. In relation to the MVCP values, for the trapezius descendens, biceps brachii, and extensor digitorum, the upwards working direction showed the highest level of muscle activation. With the flexor carpi ulnaris, the highest level of muscle activation was seen in the downwards working direction. This could be caused by a high level of radialduction. Due to the lack of grip force measurements, it is not possible to interpret the forces in relation to the EMG data and wrist postures. However, the levels of perceived exposures were also rated as very strong while working in upwards direction. Considering the effects of the combined exposures of hand–arm vibration and awkward posture in terms of muscle load and self-reported perceived exposures, the standard analysis of hand–arm vibration as a magnitude of the acceleration is insufficient. Therefore, combined exposures require further investigation to understand the influence of hand–arm posture and also to provide preventive actions.

Author Contributions: Conceptualization, methodology, software, validation, formal analysis, investigation, resources, data curation, writing—original draft preparation, writing—review and editing, visualization, supervision, project administration: N.R., T.W. and C.F. All authors have read and agreed to the published version of the manuscript.

Funding: This research received no external funding.

Informed Consent Statement: Informed consent was obtained from all subjects involved in the study.

Data Availability Statement: The data that support the findings of this study are available from the corresponding author, N.R., upon reasonable request.

Conflicts of Interest: The authors declare no conflict of interest.

References

1. Taylor, M.; Maeda, S.; Miyashita, K. An Investigation of the Effects of Drill Operator Posture on Vibration Exposure and Temporary Threshold Shift of Vibrotactile Perception Threshold. *Vibration* **2021**, *4*, 25. [CrossRef]
2. Aldien., Y.; Marcotte., P.; Rakheja., S.; Boileau., P.-E. Influence of hand-arm posture on biodynamic response of the human hand-arm exposed to zh-axis vibration. *Int. J. Ind. Ergon.* **2006**, *36*, 45–59. [CrossRef]
3. Besa, A.J.; Valero, F.J.; Suner, J.L.; Carballeria, J. Characterisation of the mechanical impedance of the human hand-arm system: The influence of vibration direction, hand-arm posture and muscle tension. *Ind. Ergon.* **2007**, *37*, 225–231. [CrossRef]
4. Hermanns, I.; Raffler, N.; Rolf, E.; Siegfried, F.; Benno, G. Simultaneous field measuring method of vibration and body posture for assessment of seated occupational driving tasks. *Int. J. Ind. Ergon.* **2008**, *38*, 255–263. [CrossRef]
5. Hansson, G.Å.; Asterland, P.; Skerfving, S. Acquisition and analysis of whole-day electromyographic field recordings. In Proceedings of the Second General SENIAM (Surface EMG for Non Invasive Assessment of Muscles) Workshop, Stockholm, Sweden, 13–14 June 1997.

proceedings

Proceeding Paper

Methods for the Laboratory Evaluation of HAV-Related Comfort in Cyclists †

Stefano Marelli and Marco Tarabini *

Department of Mechanical Engineering, Politecnico di Milano, 20136 Milan, Italy; stefano1.marelli@polimi.it
* Correspondence: marco.tarabini@polimi.it
† Presented at the 15th International Conference on Hand-Arm Vibration, Nancy, France, 6–9 June 2023.

Abstract: Cyclists are exposed to hand–arm vibration (HAV) for prolonged periods of time during training sessions and competitions. The vibration can reduce perceived comfort, thus limiting the ability of the cyclist to control the bike in endurance sessions. The study of HAV in cyclists in a controlled environment allows for comparisons between the effects of different postures, materials and technical solutions on perceived discomfort and on the vibration transmitted to specific body segments. This paper describes the experimental setup, the measurement chain and the data processing for the evaluation of bike comfort in the laboratory. The setup is based on single-axis or multiaxial shakers; the time history of the input vibration can be derived from on-field measurements for comparative analyses or can be selected from among classical stimuli for frequency response function evaluation (sine sweep or white noise). Comfort can be quantified via questionnaires; objective measurements can be derived from vibrations measured at different body locations using wearable accelerometers or laser doppler vibrometers. A case study is presented and discussed.

Keywords: hand–arm vibration; HAV; comfort; sport; cycling; measurements

Citation: Marelli, S.; Tarabini, M. Methods for the Laboratory Evaluation of HAV-Related Comfort in Cyclists. *Proceedings* **2023**, *86*, 1. https://doi.org/10.3390/proceedings2023086001

Academic Editors: Christophe Noël and Jacques Chatillon

Published: 4 April 2023

1. Introduction

Cyclists are exposed to hand–arm vibration generated by road (or track) irregularities and transmitted to the handlebars, pedals and saddle through the bike wheels, fork and frame. Vibrations limit comfort, and bike manufacturers are looking for solutions to attenuate the energy transmitted to the hands, in order to improve riding comfort. The possibility of developing diseases seems limited, though the value of A(8) (as defined in the ISO 5349-1) is usually high and the exposure time limit for a 20 km/h trip on a paved street is in the order of tens of minutes [1,2]. A few studies have evidenced possible health risks and discomfort related to cycling. Akuthota and colleagues reported the risk of developing the carpal tunnel syndrome as a result of long-distance cycling [3], while Capitani and Beer [4] indicated that several cyclists experience discomfort or pain after cycling because of inappropriate cycling posture, because of vibration or because of a combination of both factors. The problem of discomfort is particularly relevant for mountain bikers, gravel/cyclocross cyclists or during specific cobbles races [5]. Several studies have focused on laboratory experiments to reproduce cyclists' exposure to HAV. Lépine et al. [6] proposed road-simulating apparatus composed of two hydraulic shakers mounted below the bike wheels. The two shakers were actuated to provide only vertical motion to the wheels. Another study designed a test rig to measure the effects of gloves and handlebars while riding a bike [7]; the authors used the transmitted power and transmitted energy at the cyclist's hands as metrics to evaluate comfort with different grip materials. Tarabini and colleagues compared different grip materials and handlebars for motocross [8]; their experimental setup was based on an electrodynamic shaker reproducing the vertical vibration measured on a motocross bike. Vanwalleghem and colleagues proposed an instrumented seat and handlebar for comfort evaluation while riding a bike [9]. Our work aims to summarize our experience in tests performed for the evaluation of HAV-related riding comfort. This paper will focus on the experimental setup, on the identification of the vibration stimulus and on metrics for the subjective and objective evaluation of comfort.

2. Experimental Setup

We developed two setups for the comparison of different bike components. The first setup (i) allowed us to test the entire bike mounted on smart trainers/rollers, and is the preferred solution for the subjective evaluation of comfort. The plate of a 3D shaker supports the front wheel of the bike and imposes a vibration along the vertical and/or medio-lateral axes (Figure 1a). The rear wheel is mounted on commercial rollers to allow for long-term training in realistic conditions. With this setup, the cyclist's posture is determined using the bike frame dimensions; this implies that different bikes are needed to grant correct bike posture to different cyclists. The second setup (ii) allowed us to test the response of the handlebar itself, mounted on the head of a shaker using an interface that reproduces the fork head (Figure 1b). This is the preferred solution for the estimation of vibration transmissibility in different handlebars and tapes; the setup is simple but requires ad hoc measurements to ensure a realistic contact force distribution between the handlebar and the feet. The test duration is usually limited to a few minutes. In both cases, the shaker control accelerometer is located on the handlebar, in order to generate the desired vibration level at the interface with the hands.

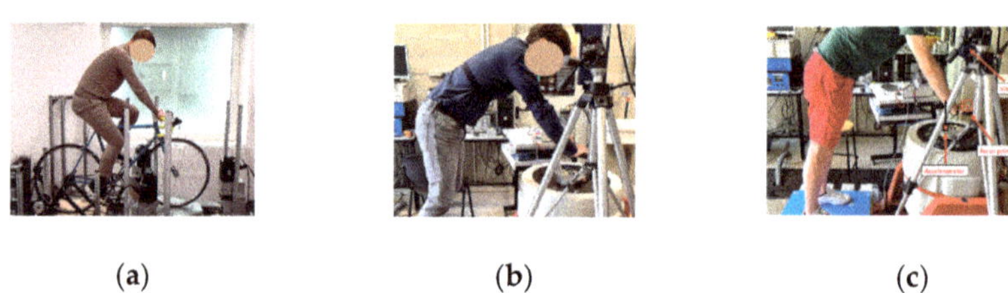

(a)   (b)   (c)

Figure 1. Pictorial view of the proposed experimental setup. (**a**) The subject is on a bike, holding the handlebar. The front wheel is placed on the shaker, while the rear wheel is fixed on a roller. (**b**) The subjects hold the handlebar mounted on a shaker through a custom interface. (**c**) Example of hand-transmitted vibration measurement chain. A vibrometer points to one knuckle of the hand, while an accelerometer measures the input vibration.

3. Vibration Stimulus

Since there are no reference vibration profiles for cycling, it is possible to adopt two approaches. The first one consists of reproducing the vibration measured on-field during a bike session. The vibration must be measured during the on-field tests at the handlebar, at the same position where the accelerometer of the shaker closed-loop control is fixed. The vibration profile depends on tests parameters such as the speed, the terrain characteristics, the tire pressure and the cyclist's anthropometric characteristics. The PSD of the vibration is then reproduced using of the two facilities described in the previous section. These kinds of experiments are focused on the evaluation of comfort using questionnaires or on the comparison of absolute (RMS) vibration transmitted to different body segments. The selection of participants among recreational or professional cyclists, with different anthropometric characteristics and ages, is strongly recommended. The second option consists of using harmonic or random stimuli; the RMS of the vibration stimulus may vary between 5 and 50 m/s^2; lower values are used to simulate urban or road cycling at low speed, while higher accelerations are meant to simulate off-road and gravel vibration. Harmonic and random stimuli are preferred for the estimation of the transfer function between the handlebar and different body segments, for the objective comparison of different materials.

4. Measurement Chain

The characteristics of the measurement chain should be derived from the ISO 8041; the input vibration on the handlebar can be detected by accelerometers with nominal sensitivity between 10 and 100 mV/ms^{-2}. The vibration transmitted to the hand and to different body segments can be measured using an accelerometer fixed either to the wrist or to the elbow using Velcro® straps or using a Laser Doppler Vibrometer pointed at a reflective tape located on a knuckle (as in Figure 1c) or on the ulnar head. When testing only the handlebar (using the second setup described in Section 2) and not the entire bike, it is important to quantify the push force or the contact pressure. The latter can be measured using capacitive or piezoresistive pressure films; qualitative measurements can be also obtained using low-cost resistive sensors, such as the FSR 408 (Interlink Electronics) or similar sensors. The push force can be measured using a triaxial force plate. The vertical component is measured via subtraction from the static weight, while the horizontal and medio-lateral components are directly measured by the force plate itself. When testing only the handlebar, the cyclist's posture has to be measured to ensure realistic testing conditions. In short-lasting tests, we typically use the Azure Kinect (Microsoft Corporation), which allows us to derive the skeleton of the cyclist after completing the tests; the parameters that we monitor are usually the wrist, elbow and shoulder angles to ensure their steadiness during the tests. The angles are computed from the joint positions and rotations given by the Kinect at a rate of 30 Hz. Alternative solutions are based on wearable sensors (such as XSens Awinda or Notch Wearable) or on optoelectronic systems (in our case, BTS Smart Evo). Wearable solutions were found to be affected by the electromagnetic field generated by the shaker when using the second setup in Section 2, mainly because the handle is close to the shaker magnets. Conversely, the time required for the setup of the markers of the optoelectronic system is high, and the use of this setup is preferred for endurance tests.

5. Metrics

Discomfort can be quantified using subjective evaluations (questionnaires) or vibration transmissibility T, expressed as a function of the vibration frequency f. In each ID tested configuration (e.g., a specific grip material or a high/low tire pressure), $T(f)$ is the ratio between the spectrum of the acceleration response $r^{ID}(f)$ and the spectrum of the vibration input $i(f)$.

$$T^{ID}(f) = \frac{r^{ID}(f)}{i(f)} \tag{1}$$

The transmissibility integral ratio (TIR) of the configuration ID can be computed as the ratio between the integral value of $T^{ID}(f)$ and the integral value of the baseline condition (e.g., the reference grip material or the nominal tire pressure) $T^{BL}(f)$

$$TIR^{ID} = \left(\int_0^{f_{Max}} T^{ID}(f)df \right) / \left(\int_0^{f_{Max}} T^{BL}(f)df \right) \tag{2}$$

Values of TIR^{ID} lower than 1 indicate better vibration attenuation of the configuration ID with respect to the baseline condition. $T^{ID}(f)$ and $T^{BL}(f)$ can be multiplied by the frequency weighting functions (for instance, w_h of ISO 5349) to give more relevance to frequencies that are more harmful or annoying for the hand–arm system. The perceived comfort can be evaluated at different time intervals using the CR100 scale proposed by Borg and Borg [10]. For research purposes, we also investigated the correlation between subjective comfort evaluation and TIR^{ID}.

6. Case Study

As an example, we describe an analysis performed to compare the effects of different tapes on comfort while riding a gravel bike. We mounted a gravel handlebar on the head of an electrodynamic shaker (LDS V830) as in Figure 1c. The stimulus was a pseudorandom signal with a PSD measured during a gravel session. The vibration at the hand was measured on the middle finger knuckle using a Polytec OFV 505 vibrometer. The input vibration was measured using a PCB Piezotronic 333B30 accelerometer. The protocol first included a measure of the baseline $T^{BL}(f)$ using no tape, with Material 1—$T^1(f)$ and with Material 2—$T^1(f)$. The protocol was repeated in three sessions (on different days). $T(f)$ was multiplied by the frequency weighting w_h; the results are summarized in Figure 2a. TIR^1 and TIR^2 are shown in Figure 2b; the results evidence that in this specific case, materials have similar vibration absorption performance. TIR variability depends on several factors, such as posture, and its variability should be carefully considered.

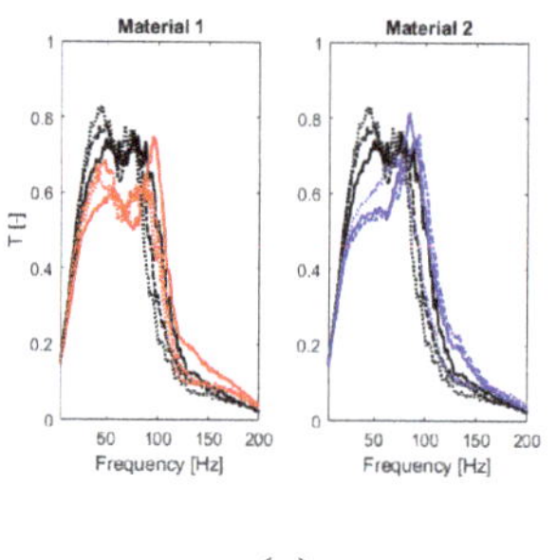
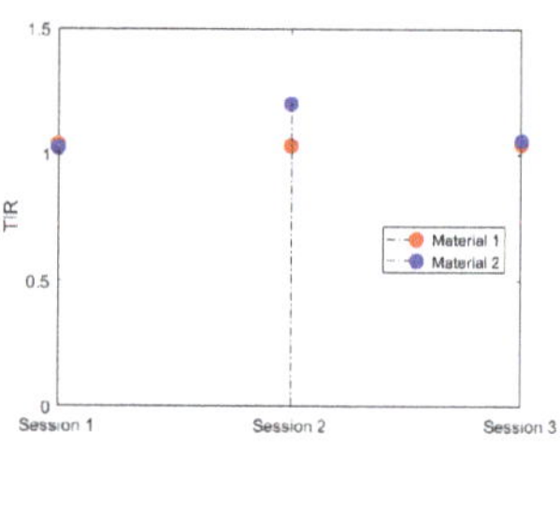

(a) **(b)**

Figure 2. (**a**) w_h weighted $T(f)$ for BL (black), material 1 (red) and material 2 (blue). Solid, dashed and dotted lines indicate the different experimental sessions. (**b**) TIR for materials 1 and 2 in the three sessions.

7. Conclusions

In this work, we described a setup and a method that enables comparison of the effectiveness of different bike materials for cycling, with the aim of quantifying their performance and possibly increasing ride comfort. Further studies are necessary to define specific frequency weighting curves to maximizing the correlation between vibration transmissibility and perceived comfort.

Author Contributions: M.T. conceptualized the work; S.M. wrote the manuscript, that was revised by M.T.; SM carried out the experiments. All authors have read and agreed to the published version of the manuscript.

Funding: SCOTT Sports SA partially supported this research.

Institutional Review Board Statement: The study was conducted in accordance with the Declaration of Helsinki. Ethical review and approval were waived owing to the lack of any type of risk for the participants.

Informed Consent Statement: Subjects involved in the study were informed about the finality of the preliminary study; informed consent was obtained by the participants.

Data Availability Statement: Data are available upon request.

Acknowledgments: The authors gratefully acknowledge Federico Stella (SCOTT Sports SA).

Conflicts of Interest: The authors declare no conflict of interest.

References

1. Chiementin, X.; Rigaut, M.; Crequy, S.; Bolaers, F.; Bertucci, W. Hand-arm vibration in cycling. *JVC/J. Vib. Control* **2013**, *19*, 2551–2560. [CrossRef]
2. Tarabini, M.; Saggin, B.; Scaccabarozzi, D. Whole-body vibration exposure in sport: Four relevant cases. *Ergonomics* **2015**, *58*, 1143–1150. [CrossRef] [PubMed]
3. Akuthota, V.; Plastaras, C.; Lindberg, K.; Tobey, J.; Press, J.; Garvan, C. The Effect of Long-Distance Bicycling on Ulnar and Median Nerves: An Electrophysiologic Evaluation of Cyclist Palsy. *Am. J. Sports Med.* **2005**, *33*, 1224–1230. [CrossRef] [PubMed]

4. Capitani, D.; Beer, S. Handlebar palsy—A compression syndrome of the deep terminal (motor) branch of the ulnar nerve in biking. *J. Neurol.* **2002**, *249*, 1441–1445. [CrossRef] [PubMed]
5. Kirkwood, L.A.; Taylor, M.D.; Ingram, L.A.; Malone, E.; Florida-James, G.D. Elite mountain bike enduro competition: A study of rider hand-arm vibration exposure. *J. Sci. Cycl.* **2019**, *8*, 18–25. [CrossRef]
6. Lépine, J.; Champoux, Y.; Drouet, J.M. Road bike comfort: On the measurement of vibrations induced to cyclist. *Sport. Eng.* **2014**, *17*, 113–122. [CrossRef]
7. Drouet, J.M.; Covill, D.; Duarte, W. On the exposure of hands to vibration in road cycling: An assessment of the effect of gloves and handlebar tape. *Proceedings* **2018**, *2*, 213. [CrossRef]
8. Tarabini, M.; Mauri, N.; Gaudio, I.; Cinquemani, S.; Moorhead, A.P.; Bongiovanni, R.; Feletti, F. Hand-arm vibration in motocross: Measurement and mitigation actions. *Muscles Ligaments Tendons J.* **2020**, *10*, 280–289. [CrossRef]
9. Vanwalleghem, J.; De Baere, I.; Loccufier, M.; Van Paepegem, W. Sensor design for outdoor racing bicycle field testing for human vibration comfort evaluation. *Meas. Sci. Technol.* **2013**, *24*, 095002. [CrossRef]
10. Borg, E.; Borg, G. A comparison of AME and CR100 for scaling perceived exertion. *Acta Psychol.* **2002**, *109*, 157–175. [CrossRef] [PubMed]

proceedings

Proceeding Paper

Comparison between the Biomechanical Responses of the Hand and Foot When Exposed to Vertical Vibration [†]

Flavia Marrone [1], Carlotta Massotti [1], Katie A. Goggins [2], Tammy R. Eger [2], Enrico Marchetti [3], Massimo Bovenzi [4] and Marco Tarabini [1,*]

[1] Department of Mechanical Engineering, Politecnico di Milano, Via Privata Giuseppe la Masa 1, 20156 Milano, Italy; flavia.marrone@polimi.com (F.M.); carlotta.massotti@polimi.com (C.M.)
[2] Centre for Research in Occupational Safety and Health, Laurentian University, Sudbury, ON P3E 2C6, Canada; kx_goggins@laurentian.ca (K.A.G.); teger@laurentian.ca (T.R.E.)
[3] INAIL, Medicine, Epidemiology, Occupational and Environmental Health Department, Via di Fontana Candida, 00040 Monte Porzio Catone, Italy; enrico2.marchetti@gmail.com
[4] Clinical Unit of Occupational Medicine, Department of Medical Sciences, University of Trieste, Via della Pietà 19, 34129 Trieste, Italy; bovenzi@units.it
* Correspondence: marco.tarabini@polimi.it; Tel.: +39-(0)2-2399 8808
† Presented at the 15th International Conference on Hand-Arm Vibration, Nancy, France, 6–9 June 2023.

Abstract: Workers can be exposed daily to foot-transmitted vibration (FTV) from standing on mobile equipment or vibrating platforms and surfaces. This results in a consistent risk of developing neurological, vascular, and musculoskeletal problems. To date, there are no international standards describing procedures with which to evaluate the health risks deriving from long-term exposure to FTV. To study the applicability of hand–arm vibration (HAV) standards to the foot, the biomechanical responses of the hand and foot in terms of the frequency response function upon varying contact conditions were compared. Results evidenced similarities between the responses of the wrist and ankle, with differences in resonance for the fingers and toes. The study confirms that HAV standards are more suitable than whole-body vibration standards for evaluating higher frequency exposure to FTV.

Keywords: hand–foot vibration similarity; foot-transmitted vibration (FTV); hand–arm vibration (HAV); vibration-induced white-foot (VIWFt)

Citation: Marrone, F.; Massotti, C.; Goggins, K.A.; Eger, T.R.; Marchetti, E.; Bovenzi, M.; Tarabini, M. Comparison between the Biomechanical Responses of the Hand and Foot When Exposed to Vertical Vibration. *Proceedings* **2023**, *86*, 34. https://doi.org/10.3390/proceedings2023086034

Academic Editors: Christophe Noël and Jacques Chatillon

Published: 18 April 2023

1. Introduction

Occupational exposure to standing foot-transmitted vibration (FTV) occurs on different means of transport (e.g., boat cabin crew members, sailors, and train operators), in manufacturing industries (operators standing on metallic surfaces close to vibrating machineries) and on heavy mining equipment (e.g., jumbo drills and bolting platforms). Prolonged standing on a vibrating floor may cause musculoskeletal disorders, motion sickness, and neurological as well as vascular diseases [1–4].

Case reports have documented the fact that occupational FTV exposure can cause vibration-induced white-foot (VIWFt), which typically manifests as Raynaud's phenomenon, with decreased blood flow, blanching, and numbness in the toes [4,5]. These vascular and neurological effects are similar to those of hand–arm vibration syndrome (HAVS), affecting workers exposed to hand-transmitted vibration (HTV) [4–10].

The International Standards Organization (ISO) has established guidelines with which to minimize the effects of occupational vibration exposure [11,12]. At the moment, the health impacts of vibration on standing and walking workers are accounted for in ISO 2631-1 [11], whose focus is dedicated to the musculoskeletal effects of whole-body vibration (WBV) on sitting, supine, and standing subjects.

The biodynamic response of the foot [13–15] and the epidemiological data on FTV [4,5] suggest that the current ISO 2631-1 [11] method of evaluating standing WBV exposure is not appropriate for evaluating vascular risks to the feet. The weighting curves used completely neglect the vascular vibration effects that cause VIWFt. For this reason, ISO 5349 [12], focused on HTV, may be more appropriate with regard to FTV for preventing health risks deriving from exposure to high-frequency vibration [9,11,13].

Considering the anatomical, biomechanical, and occupational exposure symptom similarities between the hand and foot, it seems reasonable to expect comparable frequency response functions (FRFs) between a vibration entering at the driving point and a vibration measured on the hand or on the foot.

2. Methodology

We compared the transmissibility responses of 12 paired anatomical locations of the hand [14] and foot [13,15] when exposed to vertical vibration from 10 to 150 Hz under three conditions.

Concettoni and Griffin [14] measured vibration transmissibility at 41 anatomical locations of the hand–arm system in 7 different contact conditions for 14 participants. An electrodynamic shaker imposed a random vibration to a flat plate with an RMS of 17 m/s^2 in a frequency range between 5 and 500 Hz. The transmissibility functions between acceleration at the driving point and the acceleration measured by a laser Doppler vibrometer at each anatomical point were calculated.

Goggins et al. [13,15] analyzed the vibration transmissibility at 24 anatomical locations of the foot and ankle in 3 different standing center of pressure (COP) conditions during exposure to a sine sweep of 10 to 200 Hz with a constant peak velocity of 30 mm/s. Twenty-one participants stood on a vertically vibrating platform in their natural COP position, a forward COP position, with their body weight shifted towards the toes, and in a backward COP position, with their body weight shifted into the heels. The related transmissibility curves between the input stimulus of the vibration platform and the outputs at the anatomical locations were calculated.

Paired FTV–HTV Conditions and Anatomical Locations

In the present study, similar conditions and anatomical locations between hands [14] and feet [13,15] have been compared. The considered hand–foot paired conditions are as follows (Figure 1): (i) Condition 1: whole hand on the plate compared to the natural standing COP position; (ii) Condition 2: only the fingers entirely on the plate compared to the forward COP position; and (iii) Condition 3: only the palm on the plate compared to the backward COP position.

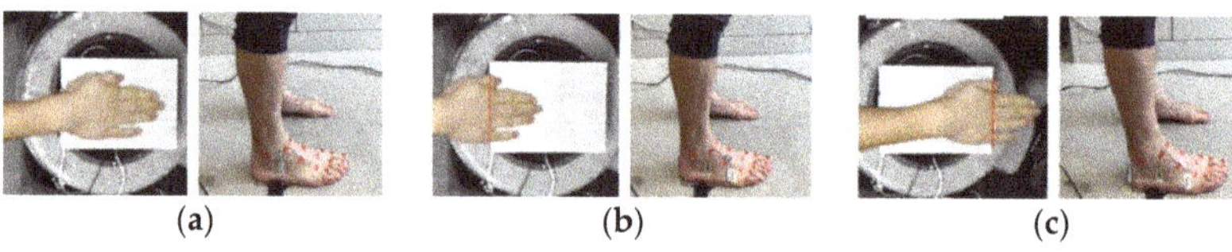

(a) (b) (c)

Figure 1. Hand–foot paired conditions. (**a**) Condition 1: whole hand–natural COP; (**b**) Condition 2: only the fingers–forward COP; and (**c**) Condition 3: only the palm–backward COP [14,15].

The considered hand–foot paired anatomical locations from both studies (Figure 2) were (i) the tips of the fingers compared to the tips of the toes (1–T1P1, 3–T2P1, 6–T3P1, 9–T4P1, and 12–T5P1), (ii) the knuckles compared to the metatarsal heads (2–T1P3, 5–T2P3, 8–T3P3, 11–T4P3, and 14–T5P3); and (iii) the wrist compared to the ankle (25–M4, 34–L4).

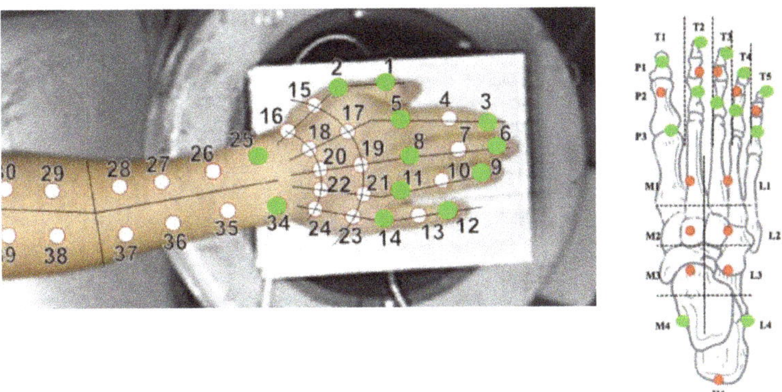

Figure 2. Anatomical locations analyzed in Concettoni and Griffin [14] as well as Goggins et al. [13,15]. Green dots represent the paired points considered in this study.

3. Results

The transmissibility response at 12 paired anatomical locations of the hand and foot were compared from 10 to 150 Hz in Conditions 1, 2, and 3 (Figure 3). In all three conditions, the transmissibility response at the wrist–ankle (i.e., 25–M4 and 34–L4) is similar, with a peak below 50 Hz and a decreasing magnitude up to 150 Hz. The transmissibility response at the finger and toe tips as well as the knuckles and metatarsal heads is similar until approximately 75 Hz. Generally, above 75 Hz, the transmissibility of the foot increases (i.e., greater than 1.5), while the hand transmissibility decreases below 1.

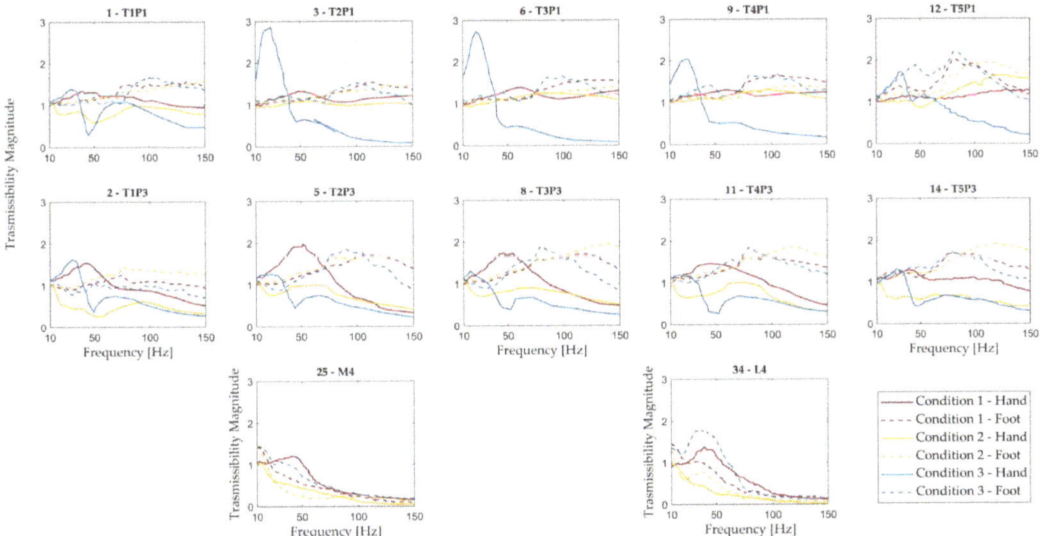

Figure 3. Hand (solid line) and paired foot (dotted line) transmissibility curves for Conditions 1 (red), 2 (yellow), and 3 (blue). Transmissibility data in Condition 3 for the wrist were not available from [14].

4. Conclusions

The comparison between the vibration transmissibility of HTV and FTV illustrates that the toes' resonance frequency (above ~80 Hz) is larger than the fingers' resonance frequency (10–60 Hz). The transmissibility response at the wrist was similar to the response at the ankle regardless of the contact condition, with the main resonance below 50 Hz in both cases. Variations in the posture, gripping/pushing forces, and usage of antivibration gloves or shoes alter the resonant frequencies.

The similarity between the vibration transmissibility of HTV and FTV suggest that using the HAV standards (i.e., ISO 5349 or its modifications) may be more appropriate for the feet; ultimately, specific standards for the evaluation of FTV exposure are necessary.

Author Contributions: Conceptualization, M.T. and M.B.; methodology, M.T.; data curation, K.A.G., T.R.E., F.M. and C.M.; writing—review and editing, F.M., M.T., K.A.G., T.R.E., M.B. and E.M.; visualization, C.M.; supervision, M.T. and M.B. All authors have read and agreed to the published version of the manuscript.

Funding: This research received no external funding.

Institutional Review Board Statement: Not applicable.

Informed Consent Statement: Not applicable.

Data Availability Statement: Not applicable.

Acknowledgments: The authors gratefully acknowledge the INAIL for supporting the Politecnico di Milano, within the BRIC funding program, throughout the years.

Conflicts of Interest: The authors declare no conflict of interest.

References

1. Musculoskeletal Disorders and Workplace Factors. A Critical Review of Epidemiologic Evidence for Work-Related Musculoskeletal Disorders of the Neck, Upper Extremity, and Low Back; U.S. Department of Health and Human Services, Public Health Service, Centers for Disease Control and Prevention, National Institute for Occupational Safety and Health: Washington, DC, USA, 1997. Available online: https://www.cdc.gov/niosh/docs/97-141/default.html (accessed on 15 January 2023).
2. Bovenzi, M. Health Effects of Mechanical Vibration. *G Ital. Med. Lav. Ergon.* **2005**, *27*, 58–64. [PubMed]
3. Tarabini, M.; Saggin, B.; Scaccabarozzi, D.; Gaviraghi, D.; Moschioni, G. Apparent Mass Distribution at the Feet of Standing Subjects Exposed to Whole-Body Vibration. *Ergonomics* **2013**, *56*, 842–855. [CrossRef] [PubMed]
4. Eger, T.; Thompson, A.; Leduc, M.; Krajnak, K.; Goggins, K.; Godwin, A.; House, R. Vibration Induced White-Feet: Overview and Field Study of Vibration Exposure and Reported Symptoms in Workers. *Work* **2014**, *47*, 101–110. [CrossRef] [PubMed]
5. Hedlund, U. Raynaud's Phenomenon of Fingers and Toes of Miners Exposed to Local and Whole-Body Vibration and Cold. *Int. Arch. Occup. Environ. Heath* **1989**, *61*, 457–461. [CrossRef] [PubMed]
6. Tingsgård, I.; Rasmussen, K. Vibration-induced white toes. *Ugeskr Laeger* **1994**, *156*, 4836–4838. [PubMed]
7. Bovenzi, M. Health Risks from Occupational Exposures to Mechanical Vibration. *Med. Lav* **2006**, *97*, 535–541. [PubMed]
8. Griffin, M.J. Measurement, Evaluation, and Assessment of Peripheral Neurological Disorders Caused by Hand-Transmitted Vibration. *Int. Arch. Occup. Environ. Health* **2008**, *81*, 559–573. [CrossRef] [PubMed]
9. Thompson, A.M.S.; House, R.; Krajnak, K.; Eger, T. Vibration-White Foot: A Case Report. *Occup. Med.* **2010**, *60*, 572–574. [CrossRef] [PubMed]
10. Shen, S.C.; House, R.A. Hand-Arm Vibration Syndrome: What Family Physicians Should Know. *Can. Fam. Physician* **2017**, *63*, 206–210. [PubMed]
11. *ISO-2631-1*; International Organization for Standardization (ISO) Mechanical Vibration and Shock—Evaluation of Human Exposure to Whole-Body Vibration—Whole-Body Vibration—Part 1: General Requirements. International Organization for Standardization (ISO): Geneva, Switzerland, 1997.
12. *ISO-5349-1*; International Organization for Standardization (ISO) Mechanical Vibration — Measurement and evaluation of human exposure to hand transmitted vibration—Part 1: General requirements. International Organization for Standardization (ISO): Geneva, Switzerland, 2001.
13. Goggins, K.A.; Tarabini, M.; Lievers, W.B.; Eger, T.R. Biomechanical Response of the Human Foot When Standing in a Natural Position While Exposed to Vertical Vibration from 10–200 Hz. *Ergonomics* **2019**, *62*, 644–656. [CrossRef] [PubMed]
14. Concettoni, E.; Griffin, M. The Apparent Mass and Mechanical Impedance of the Hand and the Transmission of Vibration to the Fingers, Hand, and Arm. *J. Sound Vib.* **2009**, *325*, 664–678. [CrossRef]
15. Goggins, K.A.; Tarabini, M.; Lievers, W.B.; Eger, T.R. Standing Centre of Pressure Alters the Vibration Transmissibility Response of the Foot. *Ergonomics* **2019**, *62*, 1202–1213. [CrossRef] [PubMed]

proceedings

Proceeding Paper

Nonlinearity of Power Absorption Curve and Hand-Arm System Physiology [†]

Enrico Marchetti [1,*], Luigi Fattorini [2], Marco Tarabini [3], Raoul Di Giovanni [1], Massimo Cavacece [4] and Angelo Tirabasso [1]

[1] INAIL, DIMEILA, Laboratory Physical Agents, 00078 Monte Porzio Catone, Italy; r.digiovanni@inail.it (R.D.G.); a.tirabasso@inail.it (A.T.)

[2] Department of Physiology and Phamracology «V. Erspamer», Università Sapienza di Roma, 00185 Rome, Italy; luigi.fattorini@uniroma1.it

[3] Department of Mechanical Engineering, Politecnico di Milano, 20156 Milano, Italy; marco.tarabini@polimi.it

[4] Department of Civil and Mechanical Engineering, University of Cassino and Southern Lazio, 03043 Cassino, Italy; cavacece@unicas.it

*   Correspondence: e.marchetti@inail.it; Tel.: +39-0694181584

†   Presented at the 15th International Conference on Hand-Arm Vibration, Nancy, France, 6–9 June 2023.

Abstract: Models of hand–arm systems (HAS) are purely mechanical. These models do not include the biological active behaviour of the system, even though it has been known since 1997 that there is a tonic vibration reflex. Since then, several authors have investigated this reflex and related it to grip force, posture and some others features of mechanical vibration power absorption. Other scholars proposed models of HAS that do not include the tonic vibration reflex and its consequences. These models, even partial models, are nonetheless effective in describing many aspects of vibration exposure. This is probably due to the complexity of the HAS, so that the confounding factors overwhelm measurements.

Keywords: power absorption; hand–arm; mechanical vibration; synchronization; tonic vibration reflex; muscular fatigue

check for updates

Citation: Marchetti, E.; Fattorini, L.; Tarabini, M.; Di Giovanni, R.; Cavacece, M.; Tirabasso, A. Nonlinearity of Power Absorption Curve and Hand-Arm System Physiology. *Proceedings* **2023**, *86*, 7. https://doi.org/10.3390/proceedings2023086007

Academic Editors: Christophe Noël and Jacques Chatillon

Published: 10 April 2023

1. Introduction

The power/energy of mechanical vibration absorbed by the hand–arm system (HAS) has been studied since the early 1970s, as this energy supposedly contributes to the eventual damage in the HAS. Mathematical models are a common way to better study the input–output relationship in complex systems but also in HAS investigations. In this regard, many authors have assessed the nonlinearity in different conditions and there are some attempts to explain this nonlinearity via the involvement of mechanical and physiological schemes. On the other hand, while studying models of the HAS, this nonlinearity is usually disregarded. The model benefits are related to the forecast of energy absorption by the HAS when exposed to a determined spectrum of mechanical vibration. Therefore, this nonlinearity should be included in models, particularly because the lack of linearity suggests the action of a physiological active mechanism which may be connected with the insurgence of vibration pathology. Nonetheless, the lack of inclusion of muscular synchronization does not affect the model's effectiveness.

This paper was written to point out this fact as well as the need to better understand nonlinearity in power absorption, taking its many features into account.

2. Hand–Arm System Vibration Biodynamics

The first biodynamical study of HAS was dated in 1972 with the work of Reynolds [1]. Radwin et al., in 1987, studied the interaction of grip force and frequency of mechanical vibrations, finding an influence of the frequency stimulus on grip force [2]. This suggested that the muscular activity alters the pattern of force production, both for a spinal response tonic vibration reflex (TVR), and for a motor drive modification involving superior motor areas. Successively, in 1994, Burstrom and Lundstrom [3] improved the measurement and definition of observables related to the biodynamic behaviour of the HAS. In their work, the authors pointed out that the main experimental conditions influencing measurements were the vibration direction, the grip force, the vibration level, the hand–arm posture and the constitution of hand and arm. The work of Martin and Park in 1997 [4] highlighted the TVR and proposed an electromyographical measurement standard for assessing the TVR influence in the normalized synchronization index SYNC, both within vibration frequency and far from it. The authors also suggested the hypothesis that synchronization could affect muscle fatigue. After this study, it is difficult to avoid considering physiology as part of the study of the HAS response to mechanical vibrations.

More recently, other works considered the relation between TVR, grip force and fatigue [5–9]. The findings of these papers were that muscular fiber's synchronization on the external vibration frequency probably increased muscular fatigue. The reason must be sought in the muscular response driven by the stretch reflex, i.e., the muscular contraction induced by the variations detected by muscular spindles and the Golgi tendon's receptors. This reflex allows the fibers that have a firing frequency near that of the vibrational one to contract more often than the other. This implies that there are some fibers that do not have rest, while there are others that not affected by the vibration. Evidently, neuromuscular systems have to satisfy two motor tasks: grip force production and TVR. For this, a capacity reduction to maintain a force level is expected, and this mechanism is the basis of muscle fatigue from a physiological point of view. The synchronization and the fatigue have been measured upon varying the posture and the grip force, while the direction and the level was fixed. Individual constitution was taken into account, comparing relative grip force to individual maximum voluntary contraction (MVC). The suggestion of the synchronization being a motion artifact was definitely rejected after the paper of Ritzman et al. [10].

The finding of this line of research is that absorption of vibration energy does affect muscular fatigue.

3. Models of HAS

Notwithstanding the lack of implementation of the aforementioned characteristics, models of HAS have been generated from 1972 [1] to the present day [9]. Models of HAS include linear and nonlinear characteristics [11] in order to mitigate vibration exposure deriving from handheld power tools [12] or from rig construction [13]. A noteworthy review of the work of NIOSH in that respect is the paper from Dong et al. [14].

Models are useful tools for the prediction of the HAS response to vibration exposure and they work efficiently in that sense, even without the implementation of physiological nonlinearity; the latter could help model muscular fatigue in the physiological definition, i.e., the inability to hold the muscular task for extended periods.

In general, the lack of the fatigue and physiological elements imply that models are representative of the HAS response for limited exposure times and for forces that are small (as percentages of the MVC).

4. Muscle Fatigue Measurement

The experiments described in this paper share the setup of another abstract—Interference of vibration exposures in the force production of the hand–arm system—and are designed to test the effects of muscular fatigue. Subjects are required to grip a handle with forces equal to 30% and 60% of MVC. Grip force is the result of two opposing components: push and pull. The subject was able to confirm both push and pull force and was required to balance it as close as possible to zero while keeping the prescribed grip force. Participants were exposed to vibration with a frequency of 30 Hz and amplitudes of 5–7.5 and 10 m/s^2. The effect of fatigue was analyzed by measuring the length of time for which participants could maintain the desired grip force.

Present results (summarised in Tables 1 and 2) show a limited influence of vibration on the time for which the subject could endure the motor task. Some subjects even improved their endurance under vibration, while some others shortened it.

Table 1. Time of endurance in minutes of a motor task of 30% of MVC upon varying the handle vibration.

	No Vib	5 m/s^2	7 m/s^2	10 m/s^2
1	4.53	5.10	4.75	5.03
2	5.02	5.04		
3	3.40	2.26		
4	4.35	5.12	4.02	3.55
5	4.48	3.40	3.38	4.70

Table 2. Time of endurance in minutes of a motor task of 60% of MVC upon varying the handle vibration.

	No Vib	5 m/s^2
1	1.28	1.35
2	2.01	1.00
4	0.45	1.14
5	2.40	2.02

5. Conclusions

The experiments did not evidence a significant effect of the vibration on the time for which participant could keep the MVC. Results are consistent with those of other laboratory works [15,16] in which muscular synchronization and muscular fatigue were not evident.

The conclusion is that there are some active mechanisms that intervene in the contraction while exposed to vibrations and that those mechanisms have to be studied in depth before we can obtain complete knowledge of muscular synchronization and fatigue. In other works, we encourage a larger collaboration between vibration experts and physiologists to overcome the current lack of specific knowledge on the effect of fatigue on the HAS response.

Author Contributions: Conceptualization, E.M., L.F. and M.T.; methodology, E.M., L.F. and M.C.; software, R.D.G.; validation, A.T., M.C. and R.D.G.; formal analysis, M.T.; investigation, M.C.; resources, E.M.; data curation, R.D.G.; writing—original draft preparation, E.M.; writing—review and editing, E.M., L.F. and M.T.; funding acquisition, E.M. All authors have read and agreed to the published version of the manuscript.

Funding: This research received no external funding.

Informed Consent Statement: Informed consent was obtained from all subjects involved in the study.

Conflicts of Interest: The authors declare no conflict of interest.

References

1. Reynolds, D.D.; Soedel, W. Dynamic response of the hand-arm system to a sinusoidal input. *J. Sound Vib.* **1972**, *21*, 339–353. [CrossRef]
2. Radwin, R.G.; Armstrong, T.J.; Chaffin, D.B. Power hand tool vibration effects on grip exertions. *Ergonomics* **1987**, *30*, 833–855. [CrossRef]
3. Burstrom, L.; Lundstrom, R. Absorption of vibration energy in the human hand and arm. *Ergonomics* **1994**, *37*, 879–890. [CrossRef] [PubMed]
4. Martin, B.J.; Park, H.S. Analysis of the tonic vibration reflex: Influence of vibration variables on motor unit synchronization and fatigue. *Eur. J. Appl. Physiol.* **1997**, *75*, 504–511. [CrossRef] [PubMed]

5. Brerro-Saby, C.; Delliaux, S.; Steinberg, J.G.; Jammes, Y. Fatigue induced changes in tonic vibration response (TVR) in humans: Relationship between electromyographic and biochemical events. *Muscle Nerve* **2008**, *38*, 1481–1489. [CrossRef] [PubMed]
6. Widia, M.; Md Dawal, Z. The effect of hand-held vibrating tools on muscle activity and grip strength. *Aust. J. Basic Appl. Sci.* **2011**, *5*, 198–211.
7. Fattorini, L.; Tirabasso, A.; Lunghi, A.; Di Giovanni, R.; Sacco, F.; Marchetti, E. Muscular forearm activation in hand-grip tasks with superimposition of mechanical vibrations. *J. Electromyogr. Kinesiol.* **2016**, *26*, 143–148. [CrossRef] [PubMed]
8. Fattorini, L.; Tirabasso, A.; Lunghi, A.; Di Giovanni, R.; Sacco, F.; Marchetti, E. Muscular synchronization and hand-arm fatigue. *Int. J. Ind. Ergon.* **2016**, *62*, 13–16. [CrossRef]
9. Cavacece, M. Incidence of Predisposing Factors on the Human Hand-arm Response with Flexed and Extended Elbow Positions of Workers Subject to Different Sources of Vibrations. *Univers. J. Public Health* **2021**, *9*, 507–519. [CrossRef]
10. Ritzmann, R.; Kramer, A.; Gruber, M.; Gollhofer, A.; Taube, W. EMG activity during whole body vibration: Motion artifacts or stretch reflexes? *Eur. J. Appl. Physiol.* **2010**, *110*, 143–151. [CrossRef] [PubMed]
11. Rakheja, S.; Gurram, R.; Gouw, G.J. Development of linear and nonlinear hand-arm vibration models using optimization and linearization techniques. *J. Biomech.* **1993**, *26*, 1253–1260. [CrossRef] [PubMed]
12. Rakheja, S.; Wu, J.Z.; Dong, R.G.; Schopper, A.W. A comparison of biodynamic models of the human hand–arm system for applications to hand-held power tools. *J. Sound Vib.* **2002**, *249*, 55–82. [CrossRef]
13. Dong, R.G.; Welcome, D.E.; Wu, J.Z.; McDowell, T.W. Development of hand-arm system models for vibrating tool analysis and test rig construction. *Noise Control Eng. J.* **2008**, *56*, 35–44. [CrossRef]
14. Dong, R.G.; Wu, J.Z.; Xu, X.S.; Welcome, D.E.; Krajnak, K. A review of hand-arm vibration studies conducted by US NIOSH since 2000. *Vibration* **2021**, *4*, 482–528. [CrossRef] [PubMed]
15. Dong, J.H.; Dong, R.G.; Rakheja, S.; Welcome, D.E.; McDowell, T.W.; Wu, J.Z. A method for analyzing absorbed power distribution in the hand and arm substructures when operating vibrating tools. *J. Sound Vib.* **2008**, *311*, 1286–1304. [CrossRef]
16. Marchetti, E.; Sisto, R.; Lunghi, A.; Sacco, F.; Sanjust, F.; Di Giovanni, R.; Botti, T.; Morgia, F.; Tirabasso, A. An investigation on the vibration transmissibility of the human elbow subjected to hand-transmitted vibration. *Int. J. Ind. Ergon.* **2017**, *62*, 82–89. [CrossRef]

Proceeding Paper

Using an Impact Wrench in Different Working Directions—An Analysis of the Individual Forces [†]

Thomas Wilzopolski *, Nastaran Raffler and Christian Freitag

Institute for Occupational Safety and Health of the German Social Accident Insurance (IFA), Alte Heerstr. 111, 53757 Sankt Augustin, Germany; nastaran.raffler@dguv.de (N.R.); christian.freitag@dguv.de (C.F.)
* Correspondence: thomas.wilzopolski@dguv.de; Tel.: +49-30-13001-3411
† Presented at the 15th International Conference on Hand-Arm Vibration, Nancy, France, 6–9 June 2023.

Abstract: When working in different directions, factors such as awkward postures can lead to different physical stresses, which can have an influence on the effects of hand–arm vibrations. In this regard, the individual forces can have an influence on the hand–arm-vibration (HAV). In this context, a force plate can be used to determine the feed force. In the case of different working directions, the measured values determined in this way can lead to misinterpretations if the force direction is not considered correctly. Within the scope of the study, screwdriving activities were carried out using an impact wrench in different working directions by 5 test subjects. In addition to the HAV, the feed force, body posture and muscle activity were recorded and evaluated. The results showed that there was a significantly different load for similar HAV.

Keywords: awkward posture; hand–arm vibration; force measuring plate; feed force; EMG

Citation: Wilzopolski, T.; Raffler, N.; Freitag, C. Using an Impact Wrench in Different Working Directions—An Analysis of the Individual Forces. *Proceedings* **2023**, *86*, 39. https://doi.org/10.3390/proceedings2023086039

Academic Editors: Christophe Noël and Jacques Chatillon

Published: 25 April 2023

1. Introduction

In order to address the occupational disease caused by HAV, vibration exposure can be measured using the acceleration values. According to DIN 45679, the coupling forces must be considered as a correction factor. However, the data of these factors are constant and do not consider the different working directions required by different postures. Therefore, the focus of this project was to investigate the influence of the working direction on the individual workload.

2. Materials and Methods

The studies were performed with 5 healthy, voluntary, right-handed male subjects (average ± standard deviation = 31 ± 4 years old; 185 ± 4 cm high and 85 ± 11 kg). The working directions, upwards (Figure 1a) and downwards (Figure 1b), were applied in a randomized order.

A height-adjustable experimental setup was used to set a basic position for the subjects, on which 12 screwdriving operations, driving 100 mm long wood screws into an oak panel, were performed with an electrical impact screwdriver in each working direction.

2.1. Force Measuring Plate and Forces

During the tests, the test subjects were supposed to stand on a force measuring plate (FMP—Figure 1d) which recorded the forces in the X- and Z-axis (F_X, F_Z).

To determine the total force (F_f) from the forces of the individual measuring axes, a vector was calculated using Formula (1).

$$F_f = \sqrt{F_X^2 + F_Z^2} \tag{1}$$

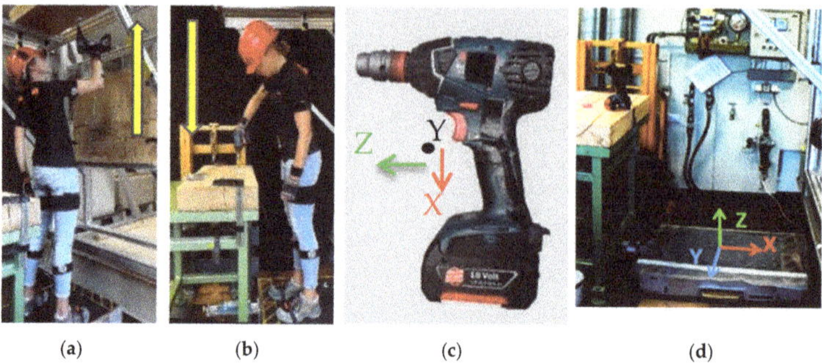

(a)   (b)   (c)   (d)

Figure 1. (**a**,**b**) Test setup for examining the two working directions, (**c**) impact wrench with accelerometer, (**d**) force measurement platform.

According to ISO 15230, the feed force (F_f) is the external force acting on the machine. In addition, the weight force of the device and the arm are defined as $F_{w\,wrench}$ and $F_{w\,arm}$. The weight force of the arm was determined using the Dortmunder model [1] as a dependence of the total weight of the individual test subjects.

Figure 2 shows the interaction of the forces for the two working directions. The total force of the subject ($F_{subject}$) is calculated for each direction following Equations (2) and (3).

$$F_{subject\,up} = F_{f\,up} + F_{w\,wrench} + F_{w\,arm} \tag{2}$$

$$F_{subject\,down} = F_{f\,down} - F_{w\,wrench} - F_{w\,arm} \tag{3}$$

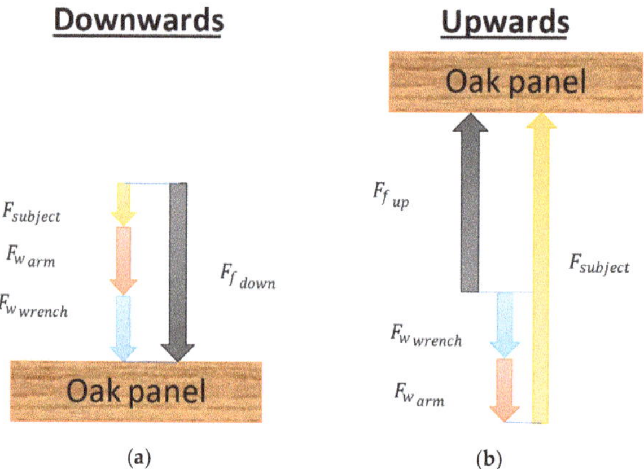

Figure 2. Interaction of forces in the working directions (**a**) upwards (**b**) downwards.

2.2. Hand–Arm Vibration

In accordance with ISO 5349 1 and 2, the HAV was recorded on the handle of the machine during the tests. For this purpose, the accelerometer was attached to the handle of the tool using cyanoacrylate, in accordance with ISO 28927 5. For the evaluation of the measurement results, the total vibration value (a_{hv}) as the sum of the frequency-weighted acceleration in three measurement axes is calculated using Equation (4).

$$a_{hv} = \sqrt{a_{hwx}^2 + a_{hwy}^2 + a_{hwz}^2} \tag{4}$$

2.3. Electromyography

To record muscle activity, surface electromyography (EMG) was performed using the Cometa Wave Plus measurement system. Sensors were placed on the skin of the subjects, over the biceps brachii and trapezius descendens, using surface electrodes. The measured values were processed according to the recommendations of Hansson et al. [2], and a relative percentage value, related to maximum voluntary contraction (MVCP), was calculated for each muscle.

2.4. Data Analysis and Statistics

The evaluation of the data was carried out in the software WIDAAN, an analysing software from the Institute for Occupational Safety and Health of the German Social Accident Insurance [3]. For this purpose, the individual screwing operations of the subjects were considered, and the individual measurement systems were synchronized with the help of video data.

3. Results & Discussion

Working downwards an a_{hv} of 5.0 ± 0.5 ms^{-2}, and working upwards a value 4.8 ± 0.6 ms^{-2}, was measured. This shows a similar workload in the different working directions.

Figure 3a shows the forces resulted from the FMP. It was shown that a significantly greater feed force was applied when working downwards compared to working upwards. However, considering the forces of the subject derived from Equations (2) and (3), the pattern of the forces is completely different.

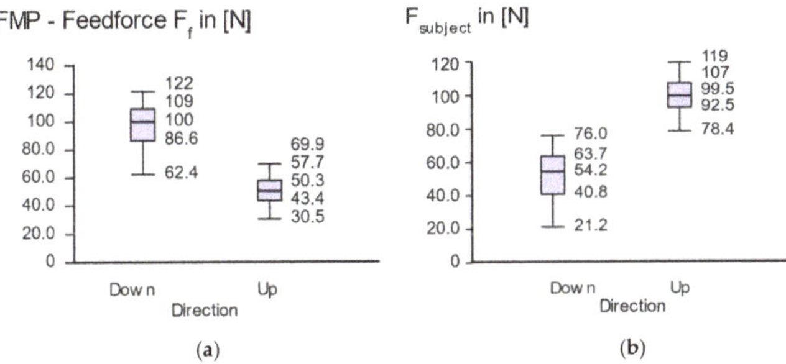

Figure 3. Measurement results of the force measuring plate (FMP) for the working directions (**a**) feed force measured with the FMP, (**b**) Fsubject considering the weight of the arm and the device.

Figure 4 shows the muscle activity values for the trapezius descendens and the biceps brachii. It indicates that there was significantly greater muscle activity when working upwards than working downwards. This also lines up very well with the calculated force $F_{subject}$ from Figure 3b.

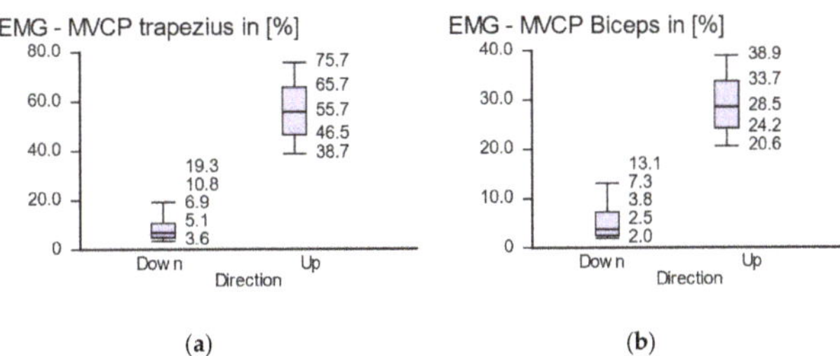

Figure 4. Results of muscle activity measurements as the percentage of the maximum voluntary contraction (MVCP) (**a**) trapezius descendens, (**b**) biceps brachii.

Although the measurement of the vibration in different working directions did not show any differences, the assessment of the forces and muscle activity indicated very different workloads for the subjects.

Summarizing these results, they highlights the fact that analysing the vibrational workload only by means of acceleration measurements overlooks many other impacting factors. Thus, an unfair and insufficient assessment of the actual workload is the consequence.

Therefore, in order to achieve a fair assessment, in addition to the correction factors for the gripping and contact forces according to DIN 45679, further factors such as the working direction must be taken into account.

Author Contributions: Conceptualization, methodology, software, validation, formal analysis, investigation, resources, data curation, writing—original draft preparation, writing—review and editing, visualization, supervision, project administration: T.W., N.R. and C.F. All authors have read and agreed to the published version of the manuscript.

Funding: This research received no external funding.

Institutional Review Board Statement: Not applicable.

Informed Consent Statement: Informed consent was obtained from all subjects involved in the study.

Data Availability Statement: The data that support the findings of this study are available from the corresponding author, T.W., upon reasonable request.

Conflicts of Interest: The authors declare no conflict of interest.

References

1. Jäger, M. The Dortmunder—Biomechanical model for quantification and assessment of the load on the lumbar spine. *J. Passeng. Car Mech. Syst. J.* **2001**, *110*, 2163–2171.
2. Hansson, G.Å.; Asterland, P.; Skerfving, S. Acquisition and analysis of whole-day electromyographic field recordings. In Proceedings of the Second General SENIAM (Surface EMG for Non Invasive Assessment of Muscles) Workshop, Stockholm, Sweden, 13–14 June 1997.
3. Hermanns, I.; Raffler, N.; Rolf, E.; Siegfried, F.; Benno, G. Simultaneous field measuring method of vibration and body posture for assessment of seated occupational driving tasks. *Int. J. Ind. Ergon.* **2008**, *38*, 255–263. [CrossRef]

Proceeding Paper

Dupuytren's Disease in Relation to the Exposure to Hand-Transmitted Vibration: A Systematic Review and Meta-Analysis [†]

Tohr Nilsson [1,*] , Jens Wahlström [1], Eirik Reierth [2] and Lage Burström [1]

1   Department of Public Health and Clinical Medicine, Faculty of Medicine, Umeå University, 90187 Umeå, Sweden; jens.wahlstrom@umu.se (J.W.); lage.burstrom@gmail.com (L.B.)
2   Science and Health Library, University of Tromsø, UiT The Arctic University of Norway, 9037 Tromsø, Norway; eirik.reierth@uit.no
*   Correspondence: tohr.nilsson@umu.se; Tel.: +46-(0)-702-140-955
†   Presented at the 15th International Conference on Hand-Arm Vibration, Nancy, France, 6–9 June 2023.

Abstract: This systematic review covering publications in the Medline and Embase databases for the period 1946 to 2020 revealed a higher prevalence of Dupuytren's disease among men exposed to vibration compared to men not exposed to vibration. The risk assessment, also considering the risk of bias, corresponded to a roughly doubled risk of Dupuytren's disease when working with vibrating machines. The supplementary meta-analysis confirmed a more than doubled risk. A possible exposure–response relation was supported by the result from the meta-analysis, which showed a doubled risk for high exposure relative to low exposure.

Keywords: Dupuytren's disease; hand–arm vibration; systematic review; meta-analysis; vibration injury

Citation: Nilsson, T.; Wahlström, J.; Reierth, E.; Burström, L. Dupuytren's Disease in Relation to the Exposure to Hand-Transmitted Vibration: A Systematic Review and Meta-Analysis. *Proceedings* **2023**, *86*, 30. https://doi.org/10.3390/proceedings2023086030

Academic Editors: Christophe Noël and Jacques Chatillon

Published: 17 April 2023

1. Introduction

Results from recent studies on workers exposed to manual work with vibrating machines has raised the question of whether there may be an association between vibration exposure and the contraction of the connective tissue in the palm and insides of the fingers. Such a fibroproliferative connective tissue disorder characterizes Dupuytren's disease. When the disease is localized to the palmo-digital fascia (aponeurosis) that lies between the dermis and flexor tendons of the hand, the result may include the formation of nodules and cords, whereupon the finger gradually loses extension, resulting in the irreversible flexion contracture of the digit, a condition called Dupuytren´s contracture. There is currently no evidence-based systematic overview of the relation between hand–arm vibration exposure and Dupuytren's disease/contracture, where vibration exposure levels were compared and for which vulnerability in terms of the predisposition and Dupuytren's diathesis were controlled.

The following systematic review aims to specifically study the risk of Dupuytren's disease in relation to the exposure to hand-transmitted vibration. Moreover, the aim was to estimate the magnitude of such an association using statistical synthesis (meta-analysis).

2. Materials and Methods

The following systematic literature review followed the PRISMA method [1], including a literature search, relevance assessment, assessment of the risk of bias, a descriptive synthesis of the results with a qualitative descriptive evidence evaluation (narrative synthesis) as well as a limited statistical synthesis (meta-analysis) for hand–arm transmitted vibration exposure and Dupuytren's disease.

In this report, vibration exposure was categorized into the following four levels:

- Level 1 = Estimate of the total number of years during the working life that involved exposure to vibrations from working with vibrating handheld machines;
- Level 2 = Estimate of the number of hours per day of exposure to vibrations from work with vibrating hand-held machines;
- Level 3 = Assessment of the daily vibration exposure through a combination of the daily exposure time and measured vibration level on the used machines;
- Level 4 = Assessment of the cumulative total vibration exposure over the entire working life through a combination of the daily exposure time, the measured vibration level of the used machines, and the number of years in which the various exposures occurred.

The review process was based on articles found in systematic database-based literature searches. The searches covered Ovid MEDLINE®, including online searching for articles under indexed and nonindexed references, from 1946 to July 2020, as well as the Embase® Classic and Embase® databases, from 1947 to July 2020. Established MESH search terms were used for MEDLINE® and terms from the search list for Emtree for EMBASE®, as well as a free text word search. An additional manual search was carried out on the reviews and reference lists of the identified articles. During the systematic review, repeated manual updates on PubMed® ensured that the literature up to 31 December 2020 was included.

To be included in the first selection, the articles were required to contain information about Dupuytren's disease or Dupuytren's contracture and include information on exposure to hand–arm transmitted vibrations. Duplicates and articles that did not meet the inclusion criteria were excluded manually, as well as articles published in a language other than English, and those lacking information on exposure to hand-transmitted vibrations.

The assessment of the risk of bias (lack of reliability) followed a predefined protocol with established criteria for: a. diagnostic reliability, b. exposure assessment reliability, and c. the methodological reliability and scientific quality of the studies. These were summed for the measure of quality, where high numerical values indicated a low risk of bias or high reliability (the details are presented in a full open-access report, in Swedish https://hdl.handle.net/2077/71843 (accessed on 30 May 2022)) [2]).

All meta-analyses were conducted with the statistical program Comprehensive Meta-analysis. Studies that reported a relative risk (Odds Ratio) were included in the meta-analyses, as well as studies that presented data that allowed calculation of an unadjusted odds ratio.

3. Results

The search provided a total of 75 references. After removal of duplicates, 48 articles remained. In addition, 13 articles were identified by manual search. The remaining 61 articles were thereafter reviewed in their entirety to determine whether each article met the inclusion criteria. Eleven articles fulfilled the pre-established criteria for inclusion (Table 1). Fifteen were excluded due to being in a language other than English, eleven were not peer-reviewed publications, eighteen contained a lack of information on vibration, and six were not original research (reviews).

Table 1. The included studies, reference, and assessed risk of bias according to the predefined criteria, summarized for diagnosis, exposure, methodology, and summed as a total. The articles are sorted in descending order based on the summed risk of bias. A high total number implies high reliability or an expected small impact on the result from the interfering factors.

Study	Ref.	Diagnosis	Exposure	Methods	Total
Bovenzi (1994)	[3]	5 *	7	6	18
Morelli et al. (2017)	[4]	7 **	1	8	16
Haines et al. (2017)	[5]	5 **	2	8	15
Murinova et al. (2021)	[6]	5 **	3	6	14
Palmer et al. (2014)	[7]	2 *	7	4	13

Table 1. *Cont.*

Study	Ref.	Diagnosis	Exposure	Methods	Total
Descatha et al. (2012)	[8]	5 **	1	4	10
Burke et al. (2007)	[9]	4 **	1	4	9
Lucas et al. (2008)	[10]	4 **	1	4	9
Dasgupta and Harrison (1996)	[11]	3 *	1	4	8
Descatha et al. (2014)	[12]	1 **	1	6	8
Thomas and Clarke (1992)	[13]	4 **	1	2	7

* Dupuytren´s contracture only. ** Dupuytren´s contracture or Dupuytren´s disease without information on contracture.

3.1. Meta-Analysis of the Studies of the Groups "Exposed" versus "Not Exposed" to Vibration

A risk of 2.3 (95% CI 1.6–3.3) for Dupuytren's disease was obtained when the groups exposed to vibration were compared to the unexposed reference groups (Figure 1). In the forest plot, the studies are ranked in descending order with respect to the reliability (risk of bias).

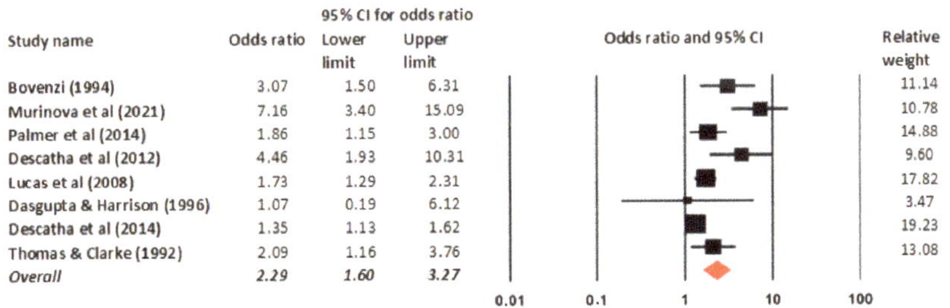

Figure 1. Statistics and the "forest plot" with the "random—effect" meta-analysis of the odds ratios of Dupuytren's disease in the vibration-exposed and non-vibration-exposed groups. The studies [3–13] are sorted in order from the highest to the lowest reliability score, according to risk of bias shown in Table 1.

3.2. Meta-Analysis of the Studies of "Low Exposure" Compared to "High Exposure"

Figure 2 shows the results of the meta-analysis for the studies that compared the risk of Dupuytren's disease between groups exposed to different levels of vibration. The analysis compared similar groups of men exposed to vibration at different exposure levels. The lowest exposure group was defined as "low exposure" and the highest exposure group as "high exposure" (Figure 2). A risk of 2.2 (95% CI 1.2–3.9) for Dupuytren's disease was obtained when the groups with "low" vibration exposure were compared to the "high"-exposure groups (Figure 2).

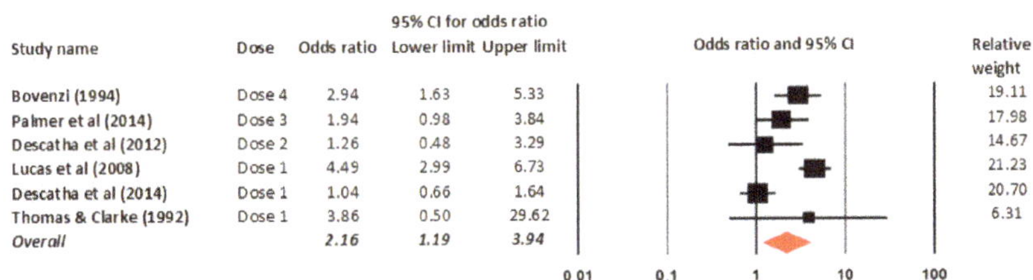

Figure 2. Statistics and the "forest plot" with "random—effect" meta-analysis of the odds ratios of Dupuytren's disease in the vibration-exposed and non-vibration-exposed groups. The studies [3,6–8,10–13] are sorted in decreasing order from the highest to the lowest reliability score, according to the rated risk of bias (Table 1).

4. Discussion

The combined results from the narrative and statistical synthesis support the conclusion that work with vibrating machines may constitute a single risk factor for Dupuytren's disease, given that the scientific basis is small, that there is an interaction between age and exposure, and that there may be individual differences in predisposition. The findings support an association between vibration and Dupuytren's disease.

Author Contributions: Conceptualization, T.N., J.W., E.R. and L.B.; methodology, T.N., J.W., E.R. and L.B.; literature search, E.R.; formal analysis, T.N., J.W., E.R. and L.B. writing—review and editing, T.N., J.W., E.R. and L.B. All authors have read and agreed to the published version of the manuscript.

Funding: This research was funded by Swedish AFA Insurance (Dnr: part of project 090326).

Data Availability Statement: Additional information is presented in reference [2].

Conflicts of Interest: The authors declare no conflict of interest.

References

1. Moher, D.; Liberati, A.; Tetzlaff, J.; Altman, G.; The Prisma Group. Preferred Reporting Items for Systematic Reviews and Meta-Analyses: The PRISMA Statement. *PLoS Med.* **2009**, *6*, e1000097. [CrossRef] [PubMed]
2. Nilsson, T.; Wahlström, J.; Reierth, E.; Burström, L. Dupuytrens sjukdom i relation till exponering för handöverförda vibrationer En Systematisk kunskapsöversikt och meta-analys. *Arb. Hälsa Vetensk. Skr.* **2022**, *56*, 1–46.
3. Bovenzi, M. Hand-arm vibration syndrome and dose-response relation for vibration induced white finger among quarry drillers and stonecarvers. Italian Study Group on Physical Hazards in the Stone Industry. *Occup Env. Med* **1994**, *51*, 603–611. [CrossRef]
4. Morelli, I.; Fraschini, G.; Banfi, A.E. Dupuytren's Disease: Predicting Factors and Associated Conditions. A Single Center Questionnaire-Based Case-Control Study. *Arch. Bone Jt. Surg.* **2017**, *5*, 384–393.
5. Haines, A.; Levis, C.; Goldsmith, C.H.; Kaur, M.; Duku, E.; Wells, R.; Walter, S.D.; Rook, C.; Stock, S.; Liss, G.; et al. Dupuytren's contracture and handwork: A case-control study. *Am. J. Ind. Med.* **2017**, *60*, 724–733. [CrossRef]
6. Murinova, L.; Perecinsky, S.; Jancova, A.; Murin, P.; Legath, L. Is Dupuytren's disease an occupational illness? *Occup. Med.* **2021**, *71*, 28–33. [CrossRef] [PubMed]
7. Palmer, K.T.; D'Angelo, S.; Syddall, H.; Griffin, M.J.; Cooper, C.; Coggon, D. Dupuytren's contracture and occupational exposure to hand-transmitted vibration. *Occup. Environ. Med.* **2014**, *71*, 241–245. [CrossRef]
8. Descatha, A.; Bodin, J.; Ha, C.; Goubault, P.; Lebreton, M.; Chastang, J.F.; Imbernon, E.; Leclerc, A.; Goldberg, M.; Roquelaure, Y. Heavy manual work, exposure to vibration and Dupuytren's disease? Results of a surveillance program for musculoskeletal disorders. *Occup. Environ. Med.* **2012**, *69*, 296–299. [CrossRef]
9. Burke, F.D.; Proud, G.; Lawson, I.J.; McGeoch, K.L.; Miles, J.N. An assessment of the effects of exposure to vibration, smoking, alcohol and diabetes on the prevalence of Dupuytren's disease in 97,537 miners. *J. Hand. Surg. Eur. Vol.* **2007**, *32*, 400–406. [CrossRef]
10. Lucas, G.; Brichet, A.; Roquelaure, Y.; Leclerc, A.; Descatha, A. Dupuytren's disease: Personal factors and occupational exposure. *Am. J. Ind. Med.* **2008**, *51*, 9–15. [CrossRef]
11. Dasgupta, A.K.; Harrison, J. Effects of vibration on the hand-arm system of miners in India. *Occup. Med.* **1996**, *46*, 71–78. [CrossRef] [PubMed]
12. Descatha, A.; Carton, M.; Mediouni, Z.; Dumontier, C.; Roquelaure, Y.; Goldberg, M.; Zins, M.; Leclerc, A. Association among work exposure, alcohol intake, smoking and Dupuytren's disease in a large cohort study (GAZEL). *BMJ Open* **2014**, *4*, e004214. [CrossRef] [PubMed]
13. Thomas, P.R.; Clarke, D. Vibration white finger and Dupuytren's contracture: Are they related? *Occup. Med.* **1992**, *42*, 155–158. [CrossRef] [PubMed]

Proceeding Paper

Radiographic Hand Osteoarthritis in Relation to Exposure to Hand-Transmitted Vibration: A Systematic Review and Meta-Analysis [†]

Tohr Nilsson [1,*], Jens Wahlström [1], Eirik Reierth [2] and Lage Burström [1]

1   Department of Public Health and Clinical Medicine, Faculty of Medicine, Umeå University, 90187 Umeå, Sweden; jens.wahlstrom@umu.se (J.W.); lage.burstrom@gmail.com (L.B.)

2   Science and Health Library, University of Tromsø, UiT The Arctic University of Norway, 9037 Tromsø, Norway; eirik.reierth@uit.no

*   Correspondence: tohr.nilsson@umu.se; Tel.: +46-(0)-702140955

†   Presented at the 15th International Conference on Hand-Arm Vibration, Nancy, France, 6–9 June 2023.

Abstract: This systematic review on radiographic hand osteoarthritis (HOA) covering publications in the databases Medline and Embase for the period 1947 to April 2021, with a final selection of 10 studies, revealed a high prevalence of hand osteoarthritis among both vibration-exposed men and non-exposed. The results show a non-significant, unadjusted risk-increase of about 50% for X-ray-diagnosed hand osteoarthritis for those who work with vibrating machinery compared to referents. The risk estimate does not provide reliable support that working with exposure from vibrating machines increases the risk of radiographic changes in the hands.

Keywords: osteoarthritis; hand osteoarthritis; hand–arm vibration; systematic review; meta-analysis; vibration injury

Citation: Nilsson, T.; Wahlström, J.; Reierth, E.; Burström, L. Radiographic Hand Osteoarthritis in Relation to Exposure to Hand-Transmitted Vibration: A Systematic Review and Meta-Analysis. *Proceedings* 2023, *86*, 31. https://doi.org/10.3390/proceedings2023086031

Academic Editors: Christophe Noël and Jacques Chatillon

Published: 17 April 2023

1. Introduction

Osteoarthritis is used as a collective term for joint failure that comes from disturbances in the balance between breakdown and new formation of the joint's various tissues (bone, cartilage, etc.). Osteoarthritis has historically been regarded as a disease of wear and tear. However, recent research shows that the disease has a complex background where a number of different causal factors work together to cause the disease. Disturbances in blood circulation, inflammatory and proinflammatory activity, mechanical stress, and trauma, as well as age-related processes, interact over time with hereditary disposition and occupational factors in the breakdown and deposits of bone and cartilage in joint structures. The results on cartilage and bone deposits are accompanied by pain, stiffness, and disability. Uneven bone turnover can cause changes in bone density with accompanying cavities (cysts) that are sometimes fluid-filled (vacuoles) and disturbed bone growth (osteophytes) or increased density (sclerosis), which can be depicted on plain film radiography (X-ray).

Early studies on workers exposed to primarily air-powered, striking machines reported injuries ("Die Presslufterkrankung") with skeletal changes in the form of bone cysts, skeletal changes in the bones of the hand, and joint osteoarthritis. The findings were deemed so unambiguous and extensive that bone loosening (malacia) of the lunate bone (Kienböck's disease) has been accepted since the 1930s as an occupational disease caused by vibration exposure and is included in the ILO's previous list of accepted occupational diseases (no. 505.01).

There is currently no recently updated evidence-based systematic review for the relationship between hand–arm vibration exposure and X-ray-diagnosed arthritis in the finger and wrist where the vibration exposure levels can be compared.

The following systematic review aims to specifically answer the question of whether X-ray-diagnosed hand osteoarthritis (HOA) is related to exposure to hand-transmitted vibrations. Moreover, the aim is to estimate the magnitude of such an association using statistical synthesis (meta-analyses).

2. Materials and Methods

This systematic literature review follows PRISMA's method [1] including a literature search, relevance assessment, assessment of risk of bias, descriptive synthesis of results with qualitative descriptive evidence evaluation (narrative synthesis), as well as limited statistical synthesis (meta-analysis) for hand–arm transmitted vibration exposure and hand osteoarthritis. The case definition of osteoarthritis is defined by radiographic markers for osteoarthritis localized to joints in the finger, metacarpal, and metacarpal bones. In this report, vibration exposure was categorized into the following four doses:

- Dose 1 = Estimate of the total number of years during the working life that involved exposure to vibrations from working with vibrating handheld machines;
- Dose 2 = Estimate of the number of hours per day that involved exposure;
- for vibrations from work with vibrating handheld machines;
- Dose 3 = Assessment of the daily vibration exposure through a combination of the daily exposure time and measured vibration level on the used machines;
- Dose 4 = Assessment of the cumulative total vibration exposure over the entire working life through a combination of daily exposure time, measured vibration level of used machines, and number of years during which the various exposures occurred.

The review process was based on articles found at systematic database-based literature searches. The searches covered Ovid MEDLINE®, including online searching for articles under indexing and non-indexed references, from 1946 until 14 April 2021, as well as the Embase® Classic and Embase® databases, from 1947 until 14 April 2021. Established MESH search terms were used for MEDLINE®, and terms from the search list for Emtree for EMBASE®, as well as free text word search. Additional manual search was carried out on reviews and reference lists of identified articles.

The assessment of risk of bias (lack of reliability) followed a predefined protocol with established criteria for a. diagnostic reliability, b. exposure assessment reliability, and c. methodological reliability and scientific quality of studies, all summed up as a sum value for quality, where high numerical values indicate low risk of bias or high reliability (details are presented in a full-paper, open-access report, in Swedish https://gupea.ub.gu.se/handle/2077/73757 (accessed on 6 October 2022) [2].

All meta-analyses were conducted with the statistical program Comprehensive Meta-analysis. Studies that reported a relative risk (odds ratio) were included in the meta-analyses as well as studies that presented data which made calculation of an unadjusted odds ratio possible.

3. Results

The literature search identified a total of 66 references. After removal of duplicates, 43 references remained. In addition, 48 articles were identified after manual review of reference lists in the overview articles and the original studies. Ten articles fulfilled the pre-established criteria for inclusion (Table 1). Excluded were 20 due to a language other than English, 49 due to lack of information on vibration or outcome, and 9 due to none-original research (reviews).

In the end, 10 articles remained for the narrative synthesis. Of those, eight articles were included in the final meta-analysis (Figure 1).

Table 1. Included studies, reference, and assessed risk of bias according to predefined criteria, summarized for diagnosis, exposure, methodology, and summed up as a total. The articles are sorted in descending order based on summed up risk of bias. A high numerical number of totals implies high reliability or an expected small impact on the result from interfering factors.

Study	Ref.	Diagnosis	Exposure	Methods	Total
Kivekas et al. 1994	[3]	6	3	10	19
Bovenzi et al. 1987	[4]	7	5	4	16
Malchaire et al. 1986	[5]	6	5	4	15
Kumlin et al. 1973	[6]	4	2	2	8
Van den Bossche et al. 1984	[7]	3	3	2	8
Burke et al. 1977	[8]	4	1	2	7
Härkonen et al. 1984	[9]	4	2	5	7
Suzuki et al. 1978	[10]	4	1	2	7
Hellström and Andersen 1972	[11]	3	1	2	6
Laitinen et al. 1974	[12]	2	1	2	5

Study name	Risk ratio	95% CI for risk ratio Lower limit	Upper limit	Relative weight	A(8) m/s²
Kivekäs et al. 1994	1.68	0.76	3.69	11.38	≈ 5
Bovenzi et al. 1987	1.37	0.79	2.38	12.98	≈ 12
Malchaire et al. 1986	1.83	1.10	3.05	13.21	≈ 7.5
Kumlin et al. 1973	2.00	0.39	10.22	6.36	≈ 5
Van den Bossche et al. 1984	5.28	4.00	6.98	14.41	≈ 3.5
Härkonen et al. 1984	1.00	0.55	1.81	12.66	≈ 3.5
Suzuki et al. 1978	1.33	1.13	1.57	14.77	≈ 5
Hellström & Andersen 1972	0.73	0.53	1.01	14.24	≈ 5
Overall	**1.59**	**0.92**	**2.73**		

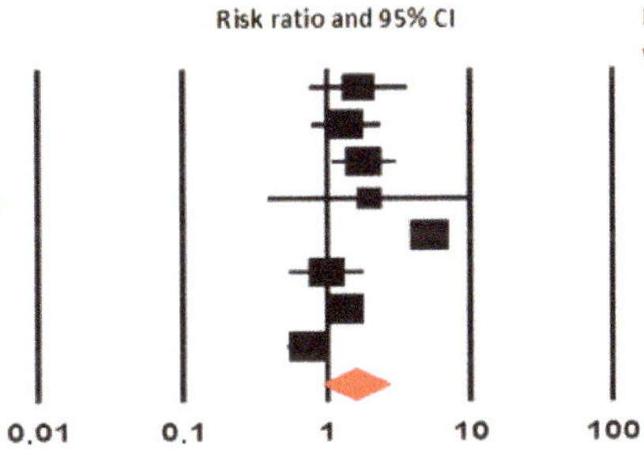

Figure 1. Statistics and "forest plot" with "random-effect" meta-analysis on prevalence ratios of hand osteoarthritis in vibration-exposed and non-vibration-exposed groups. The studies [3–12] are sorted in decreasing order from highest to lowest reliability score according to rated risk of bias (Table 1).

3.1. Prevalence of Hand Osteoarthritis in Relation to Vibration Exposure

The prevalence of hand and wrist osteoarthritis among the studies on vibration-exposed men showed a meta prevalence mean of 36 (95% CI 19 to 57), with a prevalence ranging from 11 to 83. The corresponding values for non-vibration-exposed men was 20 (95% CI 10 to 37) and 6 to 62, respectively.

3.2. Meta-Analysis of Groups Exposed versus Not Exposed to Vibration

The final meta-risk expressed as a hazard ratio (RR), not adjusted for the influence of confounders, from the eight studies included in the meta-analysis, for vibration-exposed versus non-vibration-exposed men to develop radiographically diagnosed hand and wrist osteoarthritis, was 1.59 (95% CI 0.92–2.73).

4. Discussion

This systematic review of English-language publications on hand osteoarthritis in relation to exposure to hand–arm transmitted vibrations cover the period 1947 to April 2021. The final synthesis includes 10 studies from 1972 to 1994 and is limited to only radiographic diagnosed osteoarthrosis in the hand/wrist. Radiographic manifestations of HOA occurred frequently among both those exposed to vibration and those not exposed to vibration.

An abnormal X-ray finding does not automatically imply clinical disease, nor does it have to be accompanied by symptoms or problems. Radiographs can show imaging abnormalities without clinical significance, but even insignificant X-ray changes can result in clinically serious disability. X-ray-defined osteoarthritis is more common in the population than symptomatic arthrosis and can thus lead to a dilution of risk.

The lack of high-quality studies and the lack of studies with quantified vibration dose preclude analysis of osteoarthritis in relation to different dose levels or taking a position on a possible dose–response relationship.

The crude results from the statistical synthesis (meta-analysis), regardless of the varying reliability of the studies, show a non-significant, unadjusted risk-increase of about 50% for X-ray-diagnosed hand osteoarthritis for those who work with vibrating machinery compared to referents.

5. Conclusions

Our findings do not provide reliable support that work with exposure from vibrating machines increases the risk of radiographic hand osteoarthritis. Forthcoming studies should in addition to radiographic findings entail information on symptomatic osteoarthritis.

Author Contributions: Conceptualization, T.N., J.W., E.R. and L.B.; methodology, T.N., J.W., E.R. and L.B.; literature search, E.R.; formal analysis, T.N., J.W., E.R. and L.B. writing—review and editing, T.N., J.W., E.R. and L.B. All authors have read and agreed to the published version of the manuscript.

Funding: This research was funded by Swedish AFA Insurance (Dnr: part of project 090326).

Data Availability Statement: Additional information is presented in reference [2].

Conflicts of Interest: The authors declare no conflict of interest.

References

1. Moher, D.; Liberati, A.; Tetzlaff, J.; Altman, G.; The Prisma Group. Preferred Reporting Items for Systematic Reviews and Meta-Analyses: The PRISMA Statement. *PLoS Med.* **2009**, *6*. [CrossRef] [PubMed]
2. Nilsson, T.; Wahlström, J.; Reierth, E.; Burström, L. Röntgendiagnosticerad handartros i relation till exponering för handöverförda vibrationer En systematisk litteraturöversikt med meta-analys. *Arb. Hälsa Vetensk. Skr.* **2022**, *56*, 1–53.
3. Kivekas, J.; Riihimaki, H.; Husman, K.; Hanninen, K.; Harkonen, H.; Kuusela, T.; Pekkarinen, M.; Tola, S.; Zitting, A.J. Seven-year follow-up of white-finger symptoms and radiographic wrist findings in lumberjacks and referents. *Scand J. Work Environ. Health* **1994**, *20*, 101–106. [CrossRef] [PubMed]
4. Bovenzi, M.; Fiorito, A.; Volpe, C. Bone and joint disorders in the upper extremities of chipping and grinding operators. *Int. Arch. Occup. Environ. Health* **1987**, *59*, 189–198. [CrossRef] [PubMed]
5. Malchaire, J.; Maldague, B.; Huberlant, J.M.; Croquet, F. Bone and joint changes in the wrists and elbows and their association with hand and arm vibration exposure. *Ann. Occup. Hyg.* **1986**, *30*, 461–468. [CrossRef] [PubMed]
6. Kumlin, T.; Wiikeri, M.; Sumari, P. Radiological changes in carpal and metacarpal bones and phalanges caused by chain saw vibration. *Br. J. Ind. Med.* **1973**, *30*, 71–73. [CrossRef] [PubMed]
7. Van den Bossche, J.; Lahaye, D. X ray anomalies occurring in workers exposed to vibration caused by light tools. *Br. J. Ind. Med.* **1984**, *41*, 137–141. [CrossRef] [PubMed]
8. Burke, M.J.; Fear, E.C.; Wright, V. Bone and joint changes in pneumatic drillers. *Ann. Rheum. Dis.* **1977**, *36*, 276–279. [CrossRef] [PubMed]
9. Härkönen, H.; Riihimäki, H.; Tola, S.; Mattsson, T.; Pekkarinen, M.; Zitting, A.; Husman, K. Symptoms of vibration syndrome and radiographic findings in the wrists of lumberjacks. *Br. J. Ind. Med.* **1984**, *41*, 133–136. [CrossRef] [PubMed]
10. Suzuki, K.; Takahashi, S.; Nakagawa, T. Radiological studies of the wrist joint among chain saw operating lumberjacks in Japan. *Acta Orthop. Scand.* **1978**, *49*, 464–468. [CrossRef] [PubMed]
11. Hellström, B.; Andersen, K.L. Vibration injuries in Norwegian forest workers. *Br. J. Ind. Med.* **1972**, *29*, 255–263. [CrossRef] [PubMed]
12. Laitinen, J.; Puranen, J.; Vuorinen, P. Vibration syndrome in lumbermen (working with chain saws). *J. Occup. Med.* **1974**, *16*, 552–556. [PubMed]

Proceeding Paper

Neurological Impairment from Hand–Arm Vibration Exposure [†]

Oscar Lundberg [1], Ing-Liss Bryngelsson [1] and Per Vihlborg [2,*]

1 Department of Occupational and Environmental Medicine, Faculty of Medicine and Health, Örebro University, SE 70182 Örebro, Sweden; oscar.lundberg71@gmail.com (O.L.); ing-liss.bryngelsson@regionorebrolan.se (I.-L.B.)
2 Department of Geriatrics, Faculty of Medicine and Health, Örebro University, SE 70182 Örebro, Sweden
* Correspondence: per.vihlborg@regionorebrolan.se
† Presented at the 15th International Conference on Hand-Arm Vibration, Nancy, France, 6–9 June 2023.

Abstract: The aim of this study is to investigate symtom of neurological impairment from occupational hand-arm vibration using a job exposure matrix. The result shows that paresthesia are significantly higher amongst individuals with a cumulative occupational vibration exposure over 9.08 m/s^2.

Keywords: hand-arm vibration; neuropathy; cumulative exposure; job exposure matrix

Citation: Lundberg, O.; Bryngelsson, I.-L.; Vihlborg, P. Neurological Impairment from Hand–Arm Vibration Exposure. *Proceedings* **2023**, *86*, 37. https://doi.org/10.3390/proceedings2023086037

Academic Editors: Christophe Noël and Jacques Chatillon

Published: 20 April 2023

1. Introduction

For more than 100 years, hand–arm vibration (HAV), from vibrating tools, has been reported to cause vibration white fingers (VWF), neurosensory injury and carpal tunnel syndrome (CTS) [1]. The prevalence of vibration-caused complications is difficult to estimate. Approximately 440,000 individuals in Sweden spend at least 25% of their working day using a handheld power tool [2]. This study used a job exposure matrix (JEM), a tool which provides the quantification of exposure through proxy variables of the exposure, such as exposure for specific occupations or length of employment. At a population level, this provides an accurate measurement of the exposure, but is limited by not taking the specific exposure of every individual into account more than the variables provided [3].

The aim of this study is to investigate neurological impairment in relation to HAV exposure, using the JEM.

2. Materials and Methods

The study population consisted of 1623 cases with paresthesia and an equal number of controls for each group. The controls were selected to match by sex, age, and county of residence at the case diagnosis, and were assigned by Statistics Sweden (SCB) and the National Board of Health and Welfare (SoS).

To investigate neurological impairment, we chose the diagnosis paresthesia of skin (R20.2), according to the International Classification of Diseases (ICD 10). The cases in this study were obtained from the National Board of Health and Welfare (SoS). One control for each case was then assigned by Statistics Sweden (SCB), which was selected to match by sex, age, and county of residence at the case diagnosis.

To obtain the amount of HAV exposure for every specific individual, the JEM was used. The JEM uses the employment time and occupation to estimate the individual's HAV exposure. These are through occupational codes which are specific for each occupation. The JEM could also be used to calculate the HAV exposure before the diagnosis for each individual.

Conditional logistic regression was used to calculate the odds ratio (OR) for HAV exposure, for all cases and controls.

6. Sevy, J.O.; Varacallo, M. Carpal Tunnel Syndrome. In *StatPearls*; StatPearls Publishing LLC.: Treasure Island, FL, USA, 2022. Available online: https://www.ncbi.nlm.nih.gov/books/NBK448179/ (accessed on 10 December 2022).
7. Sauni, R.; Pääkkönen, R.; Virtema, P.; Jäntti, V.; Kähönen, M.; Toppila, E.; Uitti, J. Vibration-induced white finger syndrome and carpal tunnel syndrome among Finnish metal workers. *Int. Arch. Occup. Environ. Health* **2009**, *82*, 445–453. [CrossRef] [PubMed]

Proceeding Paper

Hand-Arm Vibrations' Association with Myocardial Infarction [†]

Hans Pettersson [1,*], Claudia Lissåker [1] and Jenny Selander [2]

1 Department of Public Health and Clinical Medicine, Medical Faculty, Umea University, 90737 Umea, Sweden; claudia.lissaker@ki.se
2 Institute of Environmental Medicine, IMM, Karolinska Institute, 11365 Stockholm, Sweden; jenny.selander@ki.se
* Correspondence: hans.pettersson@umu.se; Tel.: +46-9-0785-6927
† Presented at the 15th International Conference on Hand-Arm Vibration, Nancy, France, 6–9 June 2023.

Abstract: This study found no association between exposure to hand-arm vibrations (HAV) and myocardial infarction. Data was gathered from the Swedish National Cohort on Work and Health and consists of all individuals born in Sweden from 1930 to 1990, with demographic, occupational, and MI data available between 1968 and 2017. All workers in Sweden with an occupational code between 1985 and 2013 were matched to the job-exposure matrix on occupational exposures. The model was adjusted for demographic data and other occupational exposures. The hazard ratio with a 95% confidence interval was 1.01 (0.92–1.11) for those exposed above the daily equivalent HAV level of 5 m/s^2.

Keywords: hand-arm vibration; myocardial infraction; job-exposure matrix

1. Introduction

Cardiovascular disease may be deadly or greatly affect people's lives, and one of the most important subgroups is ischemic heart disease, where myocardial infarction (MI) is the most common diagnosis. Many occupational exposures may affect the risk of MI, but less is known of Hand-Arm Vibrations (HAV) effect on this disease [1,2]. Experimental studies have found acute effects of HAV on heart rate variability [3]. Epidemiological studies on HAV effects on ischemic heart disease and myocardial infarction are few [4,5]. Miners exposed to HAV had a higher prevalence of ischemic heart disease and an increased risk of MI when exposed to HAV and WBV [4,5]. The SWE-JEM project has developed job-exposure matrices (JEM) on common occupational exposures, including HAV, to be used for studies on HAV exposure and its effect on the risk of myocardial infarction. This paper studied the association between HAV exposure and the risk of first-time myocardial infarction, with adjustment for other occupational exposures.

2. Materials and Methods

This study gathered data from the Swedish National Cohort on Work and Health (SNOW), created using Swedish registers. SNOW consists of all individuals born in Sweden from 1930 to 1990 and living in Sweden between 1968 and 2017, with demographic, occupational, and MI data available between 1968 and 2017. All workers in Sweden with an occupational code between 1985 and 2013 were matched to job-exposure matrix on HAV and other occupational exposures known to affect the risk of MI. Every participant's occupation has been coded according to the occupational classifications of the National Labour Market Board (Arbetsmarknadsstyrelsens yrkesklassificering). The occupational classification code is built on the International Standard Classification of Occupations (ISCO-88-code system) and has been described elsewhere [6]. The classification of occupations used was FOB 1980, 1985, 1990 (FOB 80 and FOB 85), and SSYK 96 and SSYK 12. Each code for every classification was given the calculated eight hour daily equivalent HAV exposure level A(8). The A(8) value was calculated from the HAV exposure from hand-held vibrating machines used by workers in each occupation and the daily duration of operating each machine according to international standard ISO 5349-1. All HAV exposure levels were calculated or gathered from earlier measurements and calculations from vibration databases, measurement reports, and scientific articles (N = 90). The A(8) exposure was categorised as 0 m/s^2 or no HAV exposure, above 0 up to 1 m/s^2, above 1 up to 2.5 m/s^2, above 2.5 up to 5 m/s^2 and above 5 m/s^2. In the SWE-JEM project, a JEM for occupational noise, demands, decision authority, physical workload index, and chemical/particle exposures has been constructed and used in the analysis. The chemical and particle exposures included were carbon monoxide, diesel exhaust, oil mist, polycyclic aromatic hydrocarbons, pulp and paper particles, silica, and welding fumes. Information on MI was gathered from the national patient registry and coded using the International Classification of Diseases, 7th, 8th, 9th, and 10th revisions (ICD-7, ICD-8, ICD-9, and ICD-10).

In the analysis, we first constructed a model of HAV exposure and MI, adjusting for year, age, gender, income, country of birth, and marital status. The second model also adjusted for occupational exposure to noise, demands, decision authority, physical workload index, and chemical exposure. The hazard ratio and a 95% confidence interval were calculated.

3. Results

Characteristics of the SNOW cohort in 2010 are gathered in Table 1. In 2010, the SNOW cohort had a total of 3,450,962 individuals (1,819,455 males/1,721,590 females), with an age median of 44 (51.4 males/48.6 females) included. Most individuals were born in Sweden (3,093,677, 87.4%) or the rest of Europe (7.3%). There were 1,921,575 individuals not married or in a registered partnership and 1,921,575 married or in a registered partnership. There were 157,510 males and 6208 females HAV exposed above <2.5 m/s^2.

Table 1. Characteristics of the cohort for working individuals by hand-arm vibration exposure included in the study for the year 2010.

		Hand-Arm Vibration (m/s^2) N (%)				
Descriptive Data		**Unexposed**	**>0–1**	**>1–2.5**	**>2.5–5**	**>5**
Age (median)		44	43	43	42	43
Gender	Male	1,266,881 (44.3)	154,448 (63.5)	240,616 (88.4)	155,066 (96.2)	2444 (94.4)
	Female	1,594,789 (55.7)	88,913 (36.5)	31,680 (11.6)	6062 (3.8)	146 (5.6)
Country of birth	Sweden	2,518,753 (88.0)	184,705 (75.9)	238,681 (87.7)	149,063 (92.5)	2475 (95.6)
	Nordics	72,582 (2.5)	6709 (2.8)	7703 (2.8)	3358 (2.1)	80 (3.1)
	Europe	126,111 (4.4)	23,170 (9.5)	14,058 (5.2)	5752 (3.6)	24 (0.9)
	Rest of the world	144,170 (5.0)	28,758 (11.8)	11,844 (4.4)	2955 (1.8)	11 (0.4)
Martial Status	Married or in a partnership *	1,352,696 (47.3)	100,035 (41.2)	101,424 (37.3)	59,707 (37.1)	848 (32.8)
	Not married or in a partnership *	1,505,054 (52.7)	142,950 (58.8)	170,555 (62.7)	101,278 (62.9)	1738 (67.2)

* Registered partnership.

The hazard ratio (HR) with a 95% confidence interval (CI) in the first model was above one unit for any exposure to HAV (HR 1.08–1.22) (Table 2). In the second model, also adjusted for occupational exposures, the HR and 95% CI were 0.91 (0.88–0.95) for those exposed above the action value of >2.5–5 m/s^2 and 1.01 (0.92–1.11) for those exposed above 5 m/s^2, respectively (Table 2).

Table 2. Association between occupational exposure to hand-arm vibration and first-time myocardial infarction one year later (Sweden, 1985–2013). Hazard Ratio (HR); 95% Confidence Interval (CI).

Hand-Arm Vibrations (m/s^2)	Exposed Cases	Model 1 [a]	Model 2 [b]
0	52,622	1.00 (ref)	1.00 (ref)
>0–1	6524	1.13 (1.10–1.16)	0.98 (0.96–1.01)
>1–2.5	10,471	1.21 (1.19–1.24)	1.02 (1.00–1.05)
>2.5–5.0	5258	1.08 (1.05–1.11)	0.91 (0.88–0.95)
>5.0	562	1.22 (1.12–1.33)	1.01 (0.92–1.11)

[a] year, age, gender, income, country of birth, and marital status. [b] year, age, gender, income, country of birth, marital status, noise, decision authority, physical workload, carbon monoxide, diesel exhaust, oil mist, polycyclic aromatic hydrocarbons, pulp and paper particles, silica, welding fumes.

4. Discussion

This study found no association between exposure to hand-arm vibrations (HAV) and a first-time myocardial infarction one year later. A JEM on HAV and other occupational exposures within the SWE-JEM project was used to adjust for other occupational exposures that may be associated with MI. By using a JEM, this study was able to gather a large number of individuals over several decades to study the risk of a first-time MI.

Author Contributions: Conceptualisation, H.P., C.L. and J.S.; methodology, H.P., J.S. and C.L.; software, J.S.; validation, J.S., C.L. and H.P.; formal analysis, C.L., J.S. and H.P.; investigation, H.P., C.L. and J.S.; resources, J.S.; data curation, J.S.; writing—original draft preparation, H.P. and C.L.; writing—review and editing, H.P., J.S. and C.L.; visualisation, C.L.; supervision, J.S.; project administration, J.S.; funding acquisition, J.S. All authors have read and agreed to the published version of the manuscript.

Funding: This research was funded by the AFA Insurance in Sweden (Grant number: 160361).

Institutional Review Board Statement: The study was conducted in accordance with the Declaration of Helsinki and approved by the Ethical Review Board in Stockholm (Protocol code: 2018/1298-31/2).

Informed Consent Statement: Subjects consent was waived due to the use of Swedish public register data, which allows universities in Sweden to do their own evaluation regarding consent and use of data without consent from subjects if it will benefit the citizens.

Data Availability Statement: Information and data from the job-exposure matrix used will be available from February 2023 on the Karolinska Institute homepage.

Acknowledgments: We appreciate the support of Bodil Björ at the Occupational and Environmental Medicine in the Region Västerbotten in Umeå, Sweden, for the construction of the JEM.

Conflicts of Interest: The authors declare no conflict of interest. The funders had no role in the design of the study, in the collection, analysis, or interpretation of data, in the writing of the manuscript, or in the decision to publish the results.

References

1. Skogstad, M.; Johannessen, H.A.; Tynes, T.; Mehlum, I.S.; Nordby, K.C.; Lie, A. Systematic review of the cardiovascular effects of occupational noise. *Occup. Med.* **2016**, *66*, 10–16. [CrossRef] [PubMed]
2. Theorell, T.; Jood, K.; Järvholm, L.S.; Vingård, E.; Perk, J.; Östergren, P.O.; Hall, C. A systematic review of studies in the contributions of the work environment to ischaemic heart disease development. *Eur. J. Public Health* **2016**, *26*, 470–477. [CrossRef] [PubMed]
3. Björ, B.; Burström, L.; Karlsson, M.; Nilsson, T.; Näslund, U.; Wiklund, U. Acute effects on heart rate variability when exposed to hand transmitted vibration and noise. *Int. Arch. Occup. Environ. Health* **2007**, *81*, 193–199. [CrossRef] [PubMed]
4. Tamaian, L.-D.; Cocarla, A. Occupational exposure to vibration and ischemic heart disease. *J. Occup. Health* **1998**, *40*, 73–76. [CrossRef]

5. Björ, B.; Burström, L.; Eriksson, K.; Jonsson, H.; Nathanaelsson, L.; Nilsson, T. Mortality from myocardial infarction in relation to exposure to vibration and dust among a cohort of iron-ore miners in Sweden. *Occup. Environ. Med.* **2010**, *67*, 154–158. [CrossRef] [PubMed]
6. Selander, J.; Albin, M.; Rosenhall, U.; Rylander, L.; Lewné, M.; Gustavsson, P. Maternal Occupational Exposure to Noise during Pregnancy and Hearing Dysfunction in Children: A Nationwide Prospective Cohort Study in Sweden. *Environ. Health Perspect.* **2015**, *124*, 855–860. [CrossRef] [PubMed]

Proceeding Paper

High-Frequency Vibration from Hand-Held Impact Wrenches and Propagation into Finger Tissue [†]

Snævar Leó Grétarsson *[iD] and Hans Lindell [iD]

RISE Research Institutes of Sweden, 431 53 Mölndal, Sweden; hans.lindell@ri.se
* Correspondence: snaevar.gretarsson@ri.se
† Presented at the 15th International Conference on Hand-Arm Vibration, Nancy, France, 6–9 June 2023.

Abstract: High-frequency shock-type vibration (HFV) with a frequency content mainly above 1250 Hz, e.g., from impact wrenches, is likely to cause a significant amount of vibration injuries and even hand-arm vibration syndrome. The objective of this study was to measure vibration up to 100 kHz with a Laser Doppler Vibrometer (LDV) and investigate the variation of vibration over the machine surface, the vibration propagation into finger tissue, and the vibration reduction on the finger tissue due to a foamed polymer layer. Our results showed that the vibration on the handle varies moderately and that the amplitudes are higher on the machine surface. A large proportion of the vibration is transferred into the finger tissue and thereby subjects the finger tissue to high-vibration amplitudes, but it is effectively reduced by a thin layer of foamed polymer.

Keywords: high-frequency vibration; ultravibration; hand-arm vibration syndrome; vibration injury; impact wrench; shock vibration

Citation: Grétarsson, S.L.; Lindell, H. High-Frequency Vibration from Hand-Held Impact Wrenches and Propagation into Finger Tissue. *Proceedings* **2023**, *86*, 10. https://doi.org/10.3390/proceedings2023086010

Academic Editors: Christophe Noël and Jacques Chatillon

Published: 10 April 2023

1. Introduction

The current standard for risk estimation of hand-arm vibration injury omits frequencies above 1400 Hz [1]. Research has long indicated that the risk from high-frequency vibration is underestimated by the current standard, e.g., [2]. A high prevalence of vibration injuries has been found in assembly workers that work with impact wrenches despite being classified as low exposure according to the current risk assessment [3]. In this study, two impact wrenches were measured up to 100 kHz with a Laser Doppler Vibrometer (LDV) at different locations to see the variation of high-frequency vibration. Measurements were also performed on the fingernail to investigate the vibration propagation into and through the finger to the nail surface and its reduction due to a polymer foam.

2. Method

Vibration was measured at different locations on two $^1/_2$ inch pneumatic impact wrenches with an LDV. The wrenches were run in a brake rig according to ISO 28927-2:2009. The impact wrenches used in the study were the Chicago Pneumatics 734H and the Würth DSS $^1/_2$ Superior. All sampling and data analysis was performed in MATLAB.

2.1. Measurements

The acceleration was measured at 14 positions on both impact wrenches to map the variation of acceleration over the surface. Ten positions were on the handle and three were on the upper body. The measurement locations can be seen in Figure 1.

The acceleration was also measured on the fingernail of the operator's index finger to see whether the vibration propagates through the finger. The finger was placed on position 9 on the CP 734H while the impact wrench was running, and the laser was kept in the same position to measure the vibration on the fingernail. The measurement was done with the bare finger in contact with the tool surface and with 2 mm of EPDM foam glued to the surface of the tool. The EPDM measurement was performed to investigate the potential of vibration reduction on the finger. The feed force was not measured during the test, but the operator was instructed to try to mimic normal tool griping. The measurement setup with the LDV aimed at the fingernail, as seen in Figure 2.

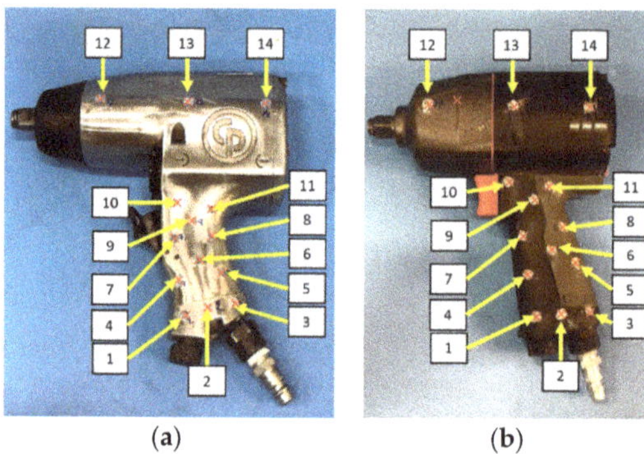

Figure 1. (**a**) Measurement positions on the CP 734H all-metal impact wrench; (**b**) measurement positions on the Würth DSS $\frac{1}{2}$" superior composite body impact wrench, except for point 12 (metal).

Figure 2. Setup for measurements on the fingernail in position 9. The green dot on the fingernail is the pilot laser that shows where the IR laser measures the velocity.

The velocity was measured using an infrared (λ = 1550 nm) LDV. Retroreflective glass beads were fixed at the measurement positions with nail polish to improve the signal quality. The Polytec Vibroflex LDV consists of a -Qtec VFX-I-160 single-point sensor head with a VFX-O-SRI short-range lens. The sensor head was connected to a Connect VXF-F-110 front end with a bandwidth of 100 kHz. The measured velocity is derived from acceleration at the front end. The acceleration signal was recorded using a National Instruments 9223 C series module with a sample rate of 1 MS/s. All positions were measured for 5 s. Position 9 was measured 3 times to investigate repeatability, which is under 10%.

2.2. Data Analysis

Currently, there is no applicable measurement standard for the evaluation of high-frequency shock vibration on hand-held tools. A proposal has been made for algorithms that quantify high-frequency shock vibration on tools [4]. The acceleration measured on the impact wrenches is highly transient. The acceleration is negligible most of the time during an impact cycle, but when the mechanism locks, high-amplitude transients occur. The magnitude of the transients varies; therefore, a method for quantifying the amplitude over time was needed. One suggestion in [4] is to define a Vibration Peak Magnitude (VPM) for a sampled acceleration signal, a_n, with N data points, as shown in Equation (1).

$$VPM(a_n) = \sqrt{\frac{\sum_{n=1}^{N} a_n^6}{\sum_{n=1}^{N} a_n^4}} \tag{1}$$

The VPM was calculated for all measurements. To investigate the influence of the upper, the VPM was calculated for the same signal with 4 bandwidths: 100 kHz, 30 kHz, 10 kHz, and 1250 Hz. The bandwidth was limited with a 6th order Butterworth low-pass filter.

3. Results

Figure 3 shows the measured signal at position 9 with a VPMVPM of 3800 m/s².

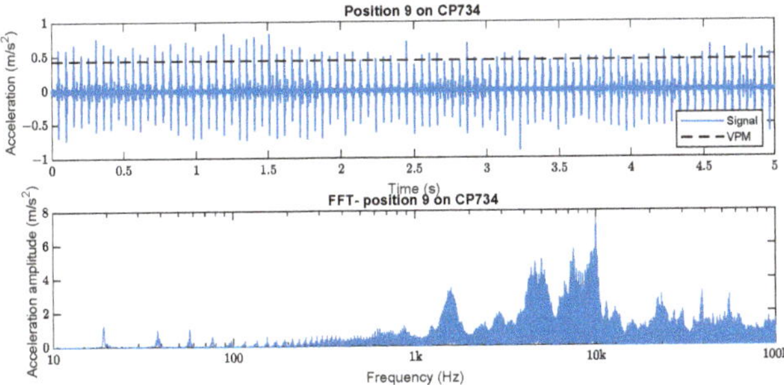

Figure 3. Top: Example of the VPM for position 9 on the CP 734H. Signal and the calculated VPM value. **Bottom:** Frequency spectra of the signal with Hann windowing.

The VPM values from different locations with different bandwidths on both impact wrenches can be seen in Figure 4. A comparison of the measured acceleration for the fingernail and the tool surface at position 9 can be seen in Figure 5. The VPM for the tool surface at position 9 was 4350 m/s². The VPM for the fingernail was 2270 m/s² with the finger directly on the tool surface at position 9. The VPM for the fingernail was 220 m/s² when the EPDM foam was between the finger and the tool at position 9.

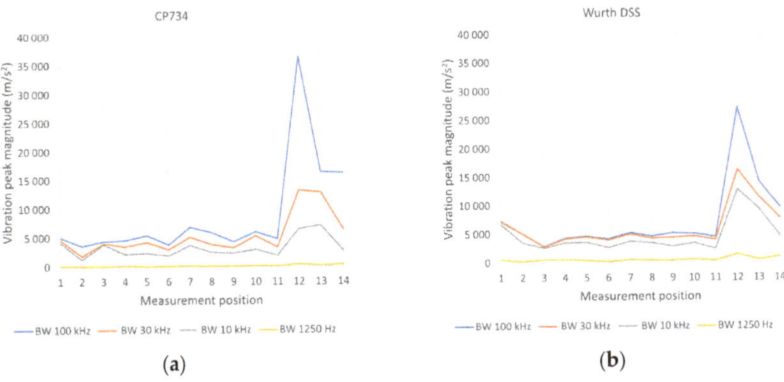

Figure 4. The vibration peak magnitude (VPM) for different measurement positions on the impact wrenches. The curves are calculated from signals that have been low-pass filtered with different cut-off frequencies. (**a**) The VPM for the CP 734H; (**b**) the VPM for the Würth DSS $^{1}/_{2}$ superior.

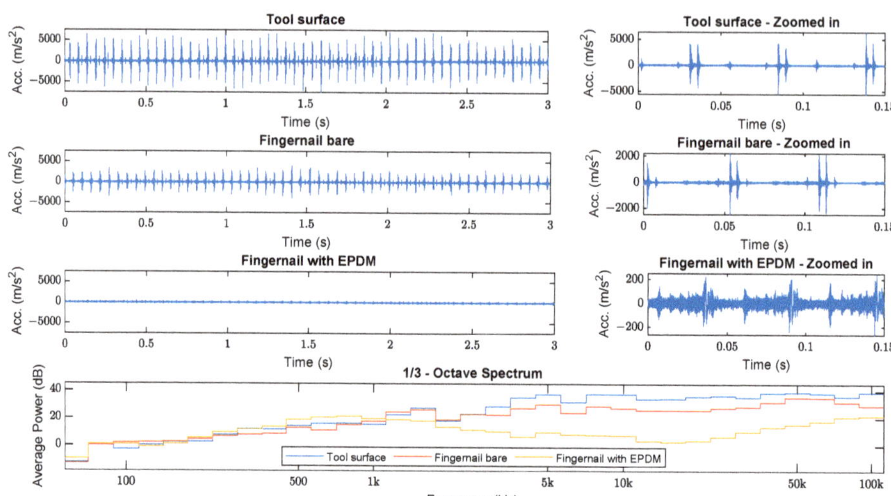

Figure 5. Comparison of measured signals at position 9 on the CP 734H. **Top row**: Acceleration for the tool surface. **Second row:** Acceleration for the fingernail when the bare finger is placed on the tool surface. **Third row:** Acceleration for the fingernail with 2 mm of EPDM foam between the tool surface and the finger. **Bottom row:** One-third octave spectrum from 20 Hz to 100 kHz for all three signals.

4. Discussion

The VPM for the handles (positions 1–11) of both impact wrenches is relatively stable around 5000 m/s^2 (Figure 4). The VPM is significantly higher (around 30,000 m/s^2) for measurement position 12, close to the main shaft. The VPM is also higher for positions 13 and 14 (i.e., the housing around the motor and mechanism).

The measurement on the fingernail showed that the VPM for the fingernail (2270 m/s^2) is around half the VPM for the tool surface below (4350 m/s^2), which indicates that high frequencies propagate through the finger. When 2 mm of foamed EPDM rubber was placed between the finger and the tool surface, the VPM was reduced by a factor of 10 (220 m/s^2), which shows that simple solutions can be used to significantly reduce exposure to high frequencies.

Author Contributions: Conceptualization, methodology, investigation, and writing, H.L. and S.L.G. Software, formal analysis, and visualization, S.L.G. All authors have read and agreed to the published version of the manuscript.

Funding: This research was funded by Vinnova, which is the Swedish Innovation Agency, and Afa Försäkring.

Informed Consent Statement: Informed consent was obtained from all subjects involved in this study.

Data Availability Statement: Not applicable.

Conflicts of Interest: The authors declare no conflict of interest. The funders had no role in the design of the study; in the collection, analyses, or interpretation of data; in the writing of the manuscript; or in the decision to publish the results.

References

1. *ISO 5349-1:2001*; Mechanical Vibration—Measurement and Evaluation of Human Exposure to Hand-Transmitted Vibration—Part 1: General Requirements. ISO (the International Organization for Standardization): Geneva, Switzerland, 2001.
2. Dandanell, R.; Engström, K. Vibration from riveting tools in the frequency range 6 Hz–10 MHz and Raynaud's phenomenon. *Scand. J. Work Environ. Health* **1986**, *12*, 338–342. [CrossRef] [PubMed]

3. Gerhardsson, L.; Ahlstrand, C.; Ersson, P.; Gustafsson, E. Vibration-induced injuries in workers exposed to transient and high frequency vibrations. *J. Occup. Med. Toxicol.* **2020**, *15*, 18. [CrossRef] [PubMed]
4. Johannisson, P.; Lindell, H. Definition and Quantification of Shock/Impact/Transient Vibrations. *arXiv* **2022**, arXiv:2211.08999.

proceedings 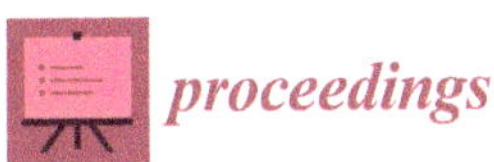

Proceeding Paper

Determination of the Number of Measurements Required for 95% Confidence in an Upper Quartile Value of Hand-Arm Vibration Measurement Using the Monte-Carlo Method [†]

Paul Pitts

Health and Safety Executive, Buxton SK17 9JN, UK; paul.pitts@hse.gov.uk

† Presented at the 15th International Conference on Hand-Arm Vibration, Nancy, France, 6–9 June 2023.

Abstract: The objective of this simulation was to determine the number of measured data sets that will provide an acceptable estimate of the upper quartile hand-arm vibration value for real use of a power tool. Monte Carlo simulations were performed based on the analysis of data sets from the HSE's hand-arm vibration database. The simulation used random uniform distribution to generate simulated machine data sets. The simulations showed that for practical measurements, a sample size of between 20 and 30 measurements is likely to achieve a reliable estimate of the upper quartile value.

Keywords: hand-arm; simulation; machine; real-use; Monte Carlo; database

Citation: Pitts, P. Determination of the Number of Measurements Required for 95% Confidence in an Upper Quartile Value of Hand-Arm Vibration Measurement Using the Monte-Carlo Method. *Proceedings* **2023**, *86*, 21. https://doi.org/10.3390/proceedings2023086021

Academic Editors: Christophe Noël and Jacques Chatillon

Published: 14 April 2023

1. Introduction

The hand-arm vibration (HAV) database operated by the Health and Safety Executive (HSE) Science Division contained (at the start of 2020) data from 11,245 in-use vibration magnitude measurements on 1636 tools. In many cases, multiple measurements had been performed on the same model of tool being used for a variety of work activities. Data from these machines have been used to assess the statistical distributions expected for real machines in real work situations.

Understanding the statistical distributions found for real machines allows the generation of large sets of randomised simulated vibration magnitude data (Monte Carlo simulations). These data sets can then be used to determine the number of measurements required to meet a target level of statistical confidence in the reported vibration value.

The objective of this simulation was to determine the number of measured data sets that will provide an estimate of the upper quartile value that is within 10% of the true value with 95% confidence.

2. Analysis of the HSE Hand-Arm Vibration Database

The HSE HAV database was interrogated in April 2020 for those machines for which more than 20 measurements have been made. The database includes 135 machines in this group. Percentile statistics of the vibration total values were calculated for each of the 135 machines. The results of these statistics for each machine were then analysed across all machines.

The number of measurements in the data sets ranged from 21 to 216 and the median number was 30, with a 75th percentile at 11.24 m/s^2 and interquartile range of 0.29 m/s^2. Figure 1 shows examples of vibration total value distributions for two machines, a random orbital sanderm and a demolition hammer.

The ratio of the interquartile range to 75th percentile value (IQR/Q75) was used as an indicator of the spread of data that is independent of absolute magnitudes. For the machines in the HSE database, the median IQR/Q75 was 0.31, with a spread from 0.10 to 0.63 for the 5th to 95th percentiles, respectively.

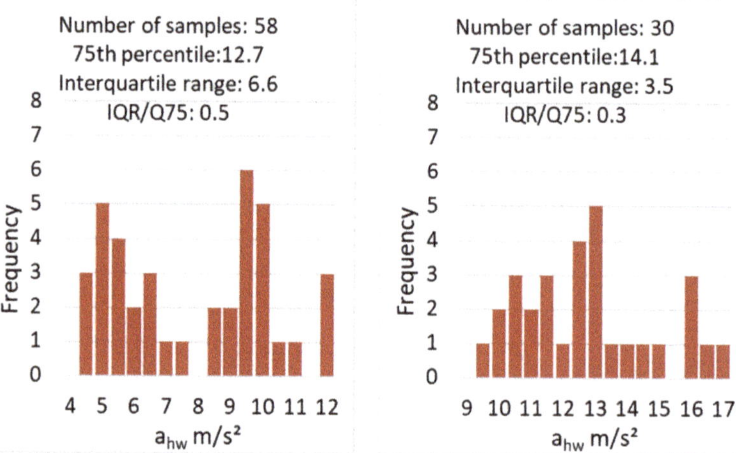

Figure 1. Examples of vibration distributions for two machine categories.

3. Monte Carlo Simulation

A Monte Carlo simulation uses random simulations of real data to analyse the likely range of potential outcomes for a real system. A uniform distribution was used, in part, for simplicity. The uniform distribution places clear bounds on the upper and lower limit of the data. Other distributions, such as log-normal, will generate many exceptionally high values that are not representative of real data.

In this analysis, many uniform random data sets were generated to represent repeated tests on a single machine. Data sets were generated containing between 5 and 50 samples. In all cases, a target upper quartile value (Q75) of 10 m/s^2 was used. Data sets were generated with different data ranges, as expressed by the ratio of IQR to the 75th percentile (IQR/Q75). The results from real machine data showed that IQR/Q75 values from 0.1 to 0.6 should be used for the simulations; values of 0.1, 0.2, 0.3, 0.4, and 0.6 were used.

For an infinite number of values, the simulation of a uniform distribution would produce a data set with a 75th percentile equal to the Q75 target value and values ranging from Q75 to 3/2 × IQR to Q75 + 1/2 × IQR. For smaller data sets, the random data sets do not appear uniform; this is illustrated in Figure 2, which shows two example distributions each of 30 simulated samples. When the distributions of multiple real data sets, such as those in Figure 1, were compared with multiple simulated data sets, such as those in Figure 2, it was concluded that the data distributions for machines in the HAV database could be represented by the simulations based on uniform data distributions.

For each combination of number of samples and the IQR/Q75 ratio, 10,000 data sets were generated. For each of the individual data sets, values for the Q75, the error in Q75 from the target Q75 value (εQ75), and the IQR were calculated.

The statistics from the 10,000 repeat sets were analysed to obtain, for each analysis combination, the 2.5th, 25th, 50th, 75th, and 97.5th percentiles of the calculated Q75 and εQ75 values, and a median value for the IQR.

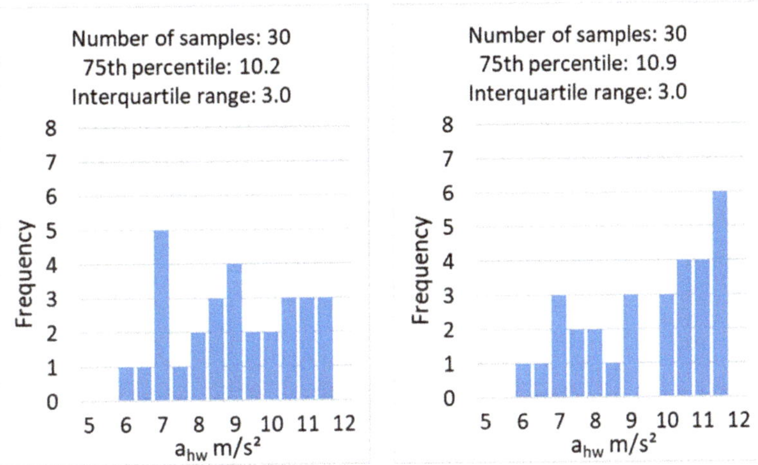

Figure 2. Examples of two simulated data distributions for 30 data samples with target 75 percentile of 10 m/s² and target interquartile range of 3 m/s².

4. Results

Figure 3 shows the results of the Monte Carlo simulation for the ratio IQR/Q75 target of 0.3. The figure shows the median error and the 95% coverage range, from 2.5% to 97.5%, for data sets ranging in size from 5 to 50 samples.

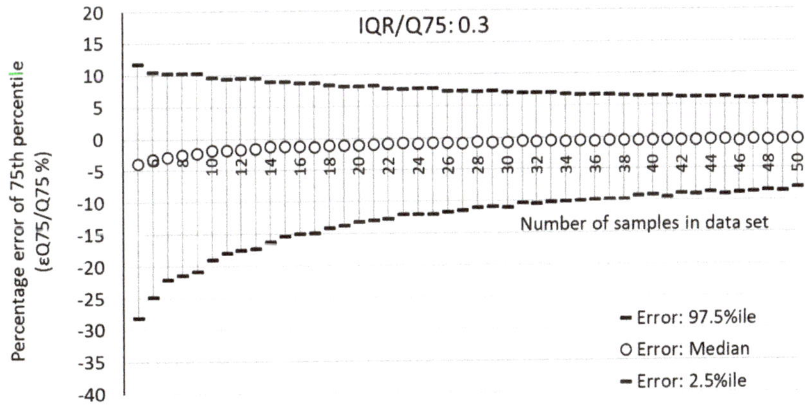

Figure 3. Comparison of analysis percentage error of the 75th percentile results for IQR/Q75 = 0.3.

Figure 3 shows that for the IRQ/Q75 = 0.3 results, a value within 20% of the true 75th percentile value is likely to be achieved (with better than 95% confidence) after ten measurements.

Due to the levelling-off of the data at higher numbers of samples, even with 10,000 data sets simulated, it is difficult to judge exactly how many samples are required to achieve a result that is just within 10% of the actual 75th percentile value. Figure 3 shows that the 10% error point is around 35 samples. However, fewer samples are still likely to achieve a result within a similar tolerance; for example, 30 measurements are likely to give a result within 11%.

Analysis was performed for five different IQR/Q75 values tested, from 0.1 to 0.6. Clearly, larger interquartile ranges led to larger error ranges. Table 1 illustrates how the error range depends on the IQR/Q75 value for sample sizes of 20 and 30 measurements.

Table 1. Comparison of percentage error of the 75th percentile results for IQR/Q75 values of 0.1 to 0.6 for 20 and 30 samples.

Number of Samples	Percentile	Percentage Error of the 75th Percentile, $\varepsilon Q75/Q75$ (%)				
		IQR/Q75 = 0.1	IQR/Q75 = 0.2	IQR/Q75 = 0.3	IQR/Q75 = 0.4	IQR/Q75 = 0.6
20	2.5th	−4.5	−8.8	−13.2	−17.7	−26.1
	50th	−0.3	−0.7	−1.1	−1.1	−2.3
	97.5th	2.7	5.4	8.1	11.1	16.6
30	2.5th	−3.5	−7.1	−11.0	−14.4	−21.8
	50th	−0.2	−0.4	−0.7	−0.8	−1.4
	97.5th	2.3	4.7	7.1	9.6	14.4

Table 1 shows that for between 20 and 30 data points, at IQR/Q75 of 0.4, the maximum error in the upper quartile value is 17.7%. For the very highly variable data, where IQR/Q75 is 0.6, the maximum likely error may reach 26.1% for 20 samples and 21.8% for 30 samples. Analysis of the HAV database showed that the highest IQR/Q75 ratio (with 95% coverage) was 0.6, so even in this likely worst case, 30 samples is highly likely to provide a result within 20% of the true value.

5. Conclusions

For measurements to provide an estimate of the upper quartile value that is within 10% of the true value with 95% confidence, around 35 measurements are required for data with the median variability in the data from the HAV database (where IQR/Q75 = 0.3).

For data sets with greater variability (IQR/Q75 = 0.4 and 0.6), between 20 and 30 measurements are required to achieve an estimate of the upper quartile value that is within 20% of the true value.

For practical measurements, a sample size of between 20 and 30 measurements is likely to achieve a reliable estimate of the true upper quartile value of real-use hand-arm vibration magnitude.

Funding: This publication and the work it describes were funded by the Health and Safety Executive (HSE), job number PHF00616-A1120. Its contents, including any opinions and/or conclusions expressed, are those of the authors alone and do not necessarily reflect HSE policy.

Institutional Review Board Statement: Not applicable.

Informed Consent Statement: Not applicable.

Data Availability Statement: Not applicable.

Conflicts of Interest: The author declares no conflict of interest.

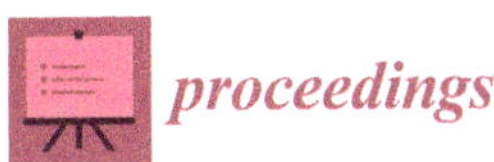

Proceeding Paper

Evaluation of Vibration Emission Values of Nailers: Can an Automatic Test Stand Be Used Instead of Human Operators? [†]

Maxime Vincent [1,*], Thomas Padois [1,2], Marc-André Gaudreau [3], Thomas Dupont [1] and Pierre Marcotte [2]

1   Département de Génie Mécanique, École de Technologie Supérieure, Montréal, QC H3C 1K3, Canada; thomas.padois@etsmtl.ca (T.P.); thomas.dupont@etsmtl.ca (T.D.)
2   Institut de Recherche Robert-Sauvé en Santé et en Sécurité du Travail, Montréal, QC H3A 3C2, Canada; pierre.marcotte@irsst.qc.ca
3   Département de Génie Mécanique, Université du Québec à Trois-Rivières, Trois-Rivières, QC G8Z 4M3, Canada; marc-andre.gaudreau@uqtr.ca
*   Correspondence: maxime.vincent.1@ens.etsmtl.ca
†   Presented at the 15th International Conference on Hand–arm Vibration, Nancy, France, 6–9 June 2023.

Abstract: To protect workers, it is necessary to characterize the noise and vibration emissions value of nailers. Standardized characterization methods exist but require three trained human operators, which leads to a dispersion in the results and a difficult implementation. An automatic test stand (ATS) was developed to characterize those values without the participation of human operators. It is proposed here that we compare the results obtained by the two methods. Preliminary results suggest that further refinement of the ATS is needed to better mimic the biodynamics of the human hand–arm and that a larger number of operators would be required for validation.

Keywords: nailers; power tools; impact vibrations; shocks; hand–arm vibration; automatic test stand

Citation: Vincent, M.; Padois, T.; Gaudreau, M.-A.; Dupont, T.; Marcotte, P. Evaluation of Vibration Emission Values of Nailers: Can an Automatic Test Stand Be Used Instead of Human Operators? *Proceedings* **2023**, *86*, 20. https://doi.org/10.3390/proceedings2023086020

Academic Editors: Christophe Noël and Jacques Chatillon

Published: 12 April 2023

1. Introduction

Portable nailers are power tools widely used in the construction industry since they permit efficient and accurate fastening of wood structures. Unfortunately, nailers also produce high levels of impact noise and vibration, which can represent a significant risk to develop hearing loss or hand–arm vibration syndrome. That is why it is important to choose and design nailers that produce low levels of noise and vibration emissions value (VEV) to prevent potential injuries. A first good step would be to simplify the methods to evaluate noise and vibration emissions value.

The ISO 28927-13:2022 standard presents a method for evaluating the VEV of fastener driving tools such as nailers [1]. The procedure requires three trained operators to fasten 50 nails each in a standardized piece of pine wood, which is particularly long and expensive. Therefore, an automatic test stand (ATS) was previously developed to simplify and to reduce the cost of VEV measurement [2,3]. Apart from the VEV characterization, the ATS was also used to determine noise emission values [2,4] and to localize nailers' noise sources [2,5]. The ATS VEV were compared with the VEV obtained with the three human operators for seven different portable nailers using the *Wh* weighted RMS acceleration (ISO 5349-1 standard [6]). However, an epidemiological study has suggested that the band-limited *hF* frequency weighting, which considers a higher frequency content, is more appropriate for assessing the risk of developing vibration-induced white finger [7]. Another study has also shown that three human operators, as specified by the ISO 10819:2013 standard [8], were insufficient to determine the vibration transmissibility of gloves [9].

In this paper, it is proposed that we compare nailers' VEV as RMS and peak *hF* weighted accelerations measured with the three human operators (ISO 28927-13:2022 standard) and with the ATS.

2. Materials and Methods

Seven different nailers were used for the study: PR1, PR2, PB1, PB2, PB3 refer to pneumatic nailers; GB1 refers to a gas (butane) one; and EB1 to an electric one. The conditions of the measurements with the human operators followed the ISO 28927-13:2022 standard, with three operators fastening 10 nails over 30 s for 5 repetitions, for a total of 50 nails by operators or 150 nails for each tested nailer.

The ATS was composed of an aluminum frame above the test setup proposed in the ISO 28927-13:2022 standard, as shown in Figure 1. The nailer handle was attached to a moving support, which consists of a greased slider restricting nailer motion to the vertical direction. A remote trigger was designed to pull the trigger of the nailer. For both ATS and human operator measurements, a triaxial accelerometer (PCB 356B20) was rigidly attached to the nailer body, as close as possible to the operator's hand. The acceleration signal was acquired with a sampling frequency of 51.2 kHz.

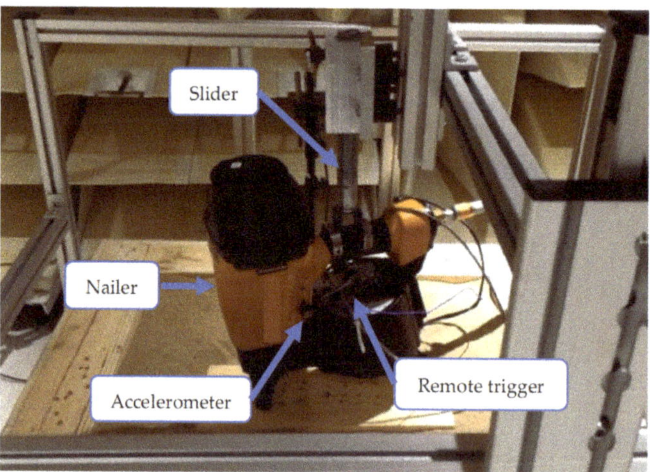

Figure 1. The automatic test stand above the ISO 28927-13:2022 test stand [2].

For the ATS, 10 nails were fastened for each nailer [2]. The acceleration signals were digitally filtered by the band limiting hF (or flat$_h$) weighting, as defined in the ISO/TS 15694 technical specification [10]. It consists in a flat band-pass filter with unity gain in the 6.3–1250 Hz frequency range. For each nailer, the 3-s RMS weighted (hF) acceleration for a single impact $a_{hF,\,3s}$, the weighted peak value $a_{hF,PEAK}$, and the crest factor CF were computed from the acceleration signals for both the three human operators and the ATS, following Equations (1)–(3). In these equations, T is the measurement period, a_{hF} is the hF weighted RMS acceleration, $a_{hF}(t)$ is the hF weighted acceleration time signal, and n is the number of single impact (nail):

$$a_{hF,\,3s} = a_{hF}\sqrt{\frac{T}{3n}}, \tag{1}$$

$$a_{hF,PEAK} = max_{0 \le t \le T}|a_{hF}(t)|, \tag{2}$$

$$CF = \frac{a_{hF,\,PEAK}}{a_{hF,3s}}. \tag{3}$$

The data obtained with the three human operators were averaged and the standard deviation was computed. For the ATS, the average and standard deviation were computed over the 10 nails. The relative difference was calculated to compare the results between the human operators and the ATS.

3. Results

The 3-s RMS, peak value, and crest factor of the hF weighted acceleration along the predominant z_h vertical direction are presented in Figure 2.

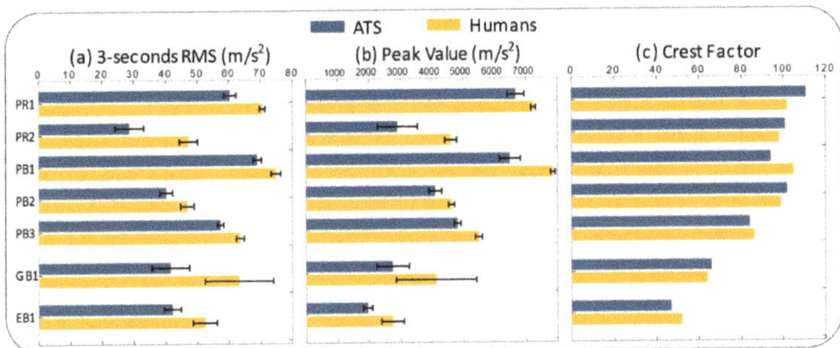

Figure 2. VEV obtained for the seven tested nailers comparing human operators and the ATS over the z_h direction: (**a**) 3-s RMS acceleration hF weighted; (**b**) peak acceleration hF weighted; and (**c**) crest factor with the mean of the three human operators and the ATS. The standard deviation is displayed as the error bar.

The results show that the 3-s RMS hF weighted acceleration and the peak value are always underestimated with the ATS. However, the CF are similar, which means the ratio between RMS and peak values is preserved. The relative differences between the 3-s RMS hF weighted acceleration obtained with the human operators and the ATS range from 8% for the PB1 nailer to 39% for the PR2 nailer. Differences for all other nailers are below 20%, except for the GB1 and PR2 nailers, which are also associated with the largest variabilities in the weighted 3-s RMS and peak accelerations. The nailer with the lowest relative difference ratio between both the 3-s RMS hF weighted acceleration and the weighted peak acceleration is the PB3 nailer, with 10% and 12%, respectively.

To further investigate those differences between the averaged human operators and the ATS, the one-third octave band spectrum of the hF weighted acceleration is computed. VEV for the PR2 and GB1 nailers, comparing the ATS with the three human operators, are presented in Figure 3. The PB3 nailer is displayed as a reference.

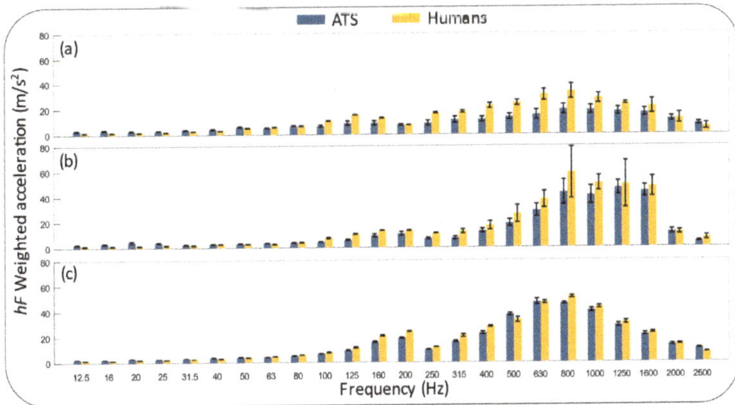

Figure 3. One-third octave band spectrum of the mean hF weighted acceleration for (**a**) PR2 nailer, (**b**) GB1 nailer, and (**c**) PB3 nailer, with the mean of the three human operators and the ATS. The standard deviation is displayed as the error bar.

The one-third octave band spectrum obtained with the PR2 nailer (Figure 3a) shows a good agreement between the ATS and human operators in the low frequency until 200 Hz. For higher frequencies, the spectrum magnitude increases for both operators and ATS as well as the difference between them. The highest values are in the 400–1600 Hz frequency range, which justify the use of the *hF* weighting. Similar observations can be made for the GB1 nailer (Figure 3b). For the human operators, it is worth noting that the error bar (i.e., the standard deviation) is very large (i.e., in the 400–1600 Hz frequency range), which demonstrates a large variability between the operators. In comparison, the PB3 nailer (Figure 3c) shows good agreement between the ATS and the human operators for the entire frequency spectra. The VEV measured with the ATS are always underestimated compared to the VEV measured with human operators. This could be attributed to the design of the ATS itself, which restricts the motion to the vertical axis only. In addition, the VEV of some nailers could be more sensitive to the human biodynamic coupling, which was not considered in the ATS design. The large intersubject variability also suggests that three operators could be insufficient to determine nailer VEV accurately.

4. Conclusions

According to the VEV comparing the ATS with the three human operators (ISO 28927-13:2022 standard), the ATS could offer a good alternative to the human operators, while it also simplifies the procedure and reduces the number of nails required to obtain the VEV. However, for two of the seven tested nailers, large differences were shown between the VEV measured with the human operators and the ATS, as well as high intersubject variabilities. These large intersubject variabilities suggest that three operators are insufficient to characterize nailer VEV. In addition, the design of the ATS, which does not consider the hand–arm biodynamics to impact vibrations, could generate VEV underestimations between the ATS and human operators. For now, more research is needed to understand the differences between the VEV measured with the ATS and with the human operators and to improve the human representativity of the ATS; an ATS with better human representativity is necessary to improve the determination of nailer VEV and noise emission values, as well as to help manufacturers build safer tools with lower levels of noise and vibration.

Author Contributions: Methodology, M.V., T.P., M.-A.G., T.D. and P.M.; writing—original draft preparation, M.V. and P.M.; writing—review and editing, M.V., T.P., M.-A.G., T.D. and P.M.; supervision, T.P., T.D. and P.M.; project administration, M.-A.G., T.D. and P.M.; funding acquisition, M.-A.G., T.D. and P.M. All authors have read and agreed to the published version of the manuscript.

Funding: The project was funded by Institut de recherche Robert-Sauvé en santé et en sécurité du travail (project # 2018-0024) and by Mitacs (project # IT 27303).

Institutional Review Board Statement: The study was conducted in accordance with the Declaration of Helsinki, and approved by the Ethics Committee of École de technologie supérieure.

Informed Consent Statement: Informed consent was obtained from all the subjects involved in the study.

Data Availability Statement: Not applicable.

Conflicts of Interest: The authors declare no conflict of interest.

References

1. *ISO 28927-13*; Hand-Held Portable Power Tools Test Methods for Evaluation of Vibration Emission Part 13: Fastener Driving Tools. International Organization for Standardization (ISO): Geneva, Switzerland, 2022.
2. Gaudreau, M.A.; Marcotte, P.; Laville, F.; Padois, T.; Boutin, J. *Cloueuses Portatives: Développement de Méthodes de Diagnostic Vibratoire et Acoustique*; IRSST: Montreal, QC, Canada, 2018; ISBN 978-2-89797-031-4.
3. Marcotte, P.; Gaudreau, M.A.; Laville, F.; Padois, T. Characterization of Nail Guns Impact Vibration. In Proceedings of the 26th International Congress on Sound and Vibration, Montréal, QC, Canada, 7–11 July 2019; pp. 1132–1138.

4.   Gaudreau, M.A.; Padois, T.; Laville, F.; Marcotte, P. Noise and Vibration Measurement of Framing Nailers: Development and Validation of a Mechanized Test Bench. In Proceedings of the 46th International Congress and Exposition on Noise Control Engineering, Hong Kong, China, 27–30 August 2017.

5.   Padois, T.; Gaudreau, M.A.; Marcotte, P.; Laville, F. Identification of Noise Sources Using a Time Domain Beamforming on Pneumatic, Gas and Electric Nail Guns. *Noise Control. Eng. J.* **2019**, *67*, 11–22. [CrossRef]

6.   *ISO 5349-1*; Mechanical Vibration—Measurement and Evaluation of Human Exposure to Hand-Transmitted Vibration—Part 1: General Requirements. International Organization for Standardization (ISO): Geneva, Switzerland, 2001.

7.   Bovenzi, M.; Pinto, I.; Picciolo, F.; Mauro, M.; Ronchese, F. Frequency Weightings of Hand-Transmitted Vibration for Predicting Vibration-Induced White Finger. *Scand. J. Work Environ. Health* **2011**, *37*, 244–252. [CrossRef] [PubMed]

8.   *ISO 10819*; Mechanical Vibration and Shock—Hand-Arm Vibration—Measurement and Evaluation of the Vibration Transmissibility of Gloves at the Palm of the Hand. International Organization for Standardization (ISO): Geneva, Switzerland, 2013.

9.   Hamouda, K.; Rakheja, S.; Dewangan, K.N.; Marcotte, P. Fingers' Vibration Transmission and Grip Strength Preservation Performance of Vibration Reducing Gloves. *Appl. Ergon.* **2018**, *66*, 121–138. [CrossRef] [PubMed]

10.   *ISO/TS 15694*; Mechanical Vibration and Shock—Measurement and Evaluation of Single Shocks Transmitted from Hand-Held and Hand-Guided Machines to the Hand-Arm System. International Organization for Standardization (ISO): Geneva, Switzerland, 2004.

Proceeding Paper

Definition and Quantification of Shock/Peak/Transient Vibration [†]

Hans Lindell [1,*], **Pontus Johannisson [2]** and **Snævar Leó Grétarsson [1]**

1   RISE Research Institutes of Sweden, 431 53 Molndal, Sweden
2   Saab, 412 76 Gothenburg, Sweden
*   Correspondence: hans.lindell@ri.se
†   Presented at the 15th International Conference on Hand-Arm Vibration, Nancy, France, 6–9 June 2023.

Abstract: Vibration injury in the hand–arm system from hand-held machines is one of the most common occupational health injuries. Machines emitting high-frequency shock vibrations, e.g., impact wrenches have since long been identified as a special risk factor. In legislative and standard texts, the terms shock, impact, peak and transient vibration are frequently used to underline the special risks associated with these kinds of vibrations. Despite this fact, in the literature there is not a mathematically stringent definition of either shock vibration or how the amplitude of the shock is defined. In this study, we suggest algorithms for definition and quantification of these terms and apply them to machine vibrations of various kinds.

Keywords: high frequency; HAVS; VPM; VSI; VSL; transient; impact; shock; vibration; ultravibration

1. Introduction

Vibration injury in the hand–arm system from hand-held machines is one of the most common occupational health injuries; it can cause severe and often chronic nerve and vascular injury to the operator. Machines emitting high-frequency shock vibrations, e.g., impact wrenches, bucking bars, chipping hammers, etc. have since long been identified as a special risk factor and the current standard for evaluation of risk is limited to frequencies below 1250 Hz [1–6]. This results in large occupational groups being exposed to harmful vibrations that are not regulated by any workers protection directives. The term ultravibration is used to define vibration with frequency above 1250 Hz, which is in analogy with ultrasound as frequencies above the human perception threshold. In legislative and standard texts the terms shock, impact, peak and transient vibration are frequently used to underline the special risk associated with these kinds of vibrations. Despite this fact, there is not a mathematically stringent definition of either shock vibration or how the amplitude of the shock is defined. To enhance the knowledge of medical effects and develop prevention measures from shock and ultra-vibration, it is of fundamental importance that they can be measured, defined and quantified.

This study suggests algorithms for definition of shock vibration and quantification of peak acceleration before applying them to measured vibrations from machines with different types of vibration characteristics. The algorithms used and the background behind them are explained in depth in Johannisson et al. [7].

Analysing shock requires sufficiently high upper frequency to cover the main energy content of the vibration. For a majority of hand-held machines, a frequency range of at least 10 kHz is suggested. Since the vibration signal is analysed in the time domain, it is recommended that the sample rate is at least five times the cut-off frequency of the low pass filter.

2. Materials and Methods

2.1. Quantifying the Peak/Shock Acceleration

A typical shock type vibration signal from an impact wrench low pass was filtered at 30 kHz (blue), 1250 Hz (purple) and with ISO 5349-1 weighting curve (yellow), as

Citation: Lindell, H.; Johannisson, P.; Grétarsson, S.L. Definition and Quantification of Shock/Peak/Transient Vibration. *Proceedings* **2023**, *86*, 29. https://doi.org//10.3390/proceedings2023086029

Academic Editors: Christophe Noël and Jacques Chatillon

Published: 14 April 2023

shown in the top graph in Figure 1; the lower graph is a zoom in on specific results. The corresponding frequency plot is seen in Figure 2.

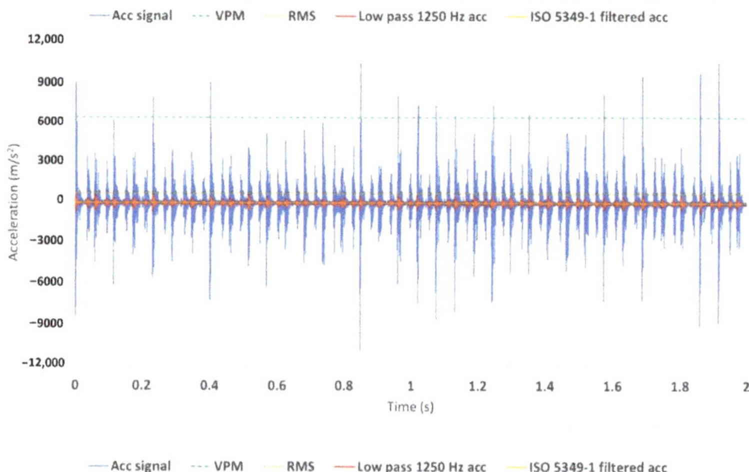

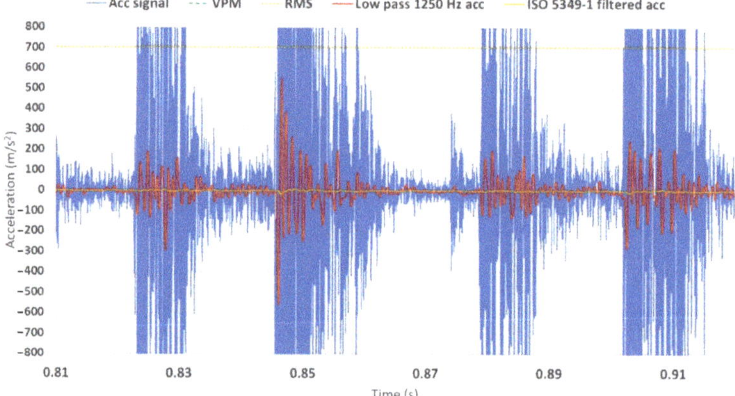

Figure 1. Acceleration from $\frac{3}{4}''$ pneumatic impact wrench, (**top**), and zoom, (**bottom**).

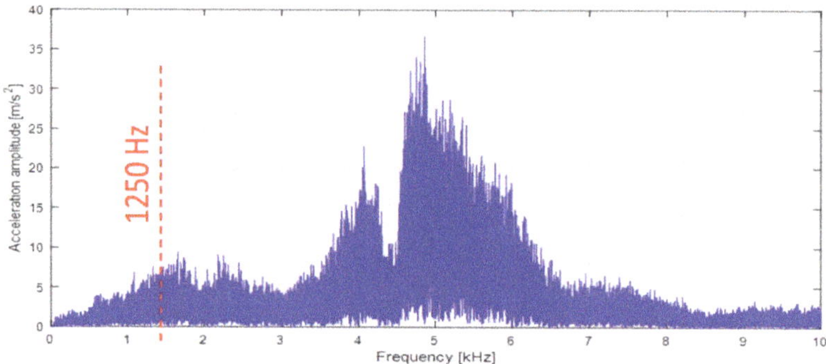

Figure 2. Acceleration frequency plot from $\frac{3}{4}''$ pneumatic impact wrench.

What could be seen is that the peak amplitude varied substantially from each impact cycle of the wrench. It could also be seen that the main frequency content was between 3 and

7 kHz; low pass filtering occurred at 1250 Hz and with ISO 5349-1 weighting, reducing the amplitude to a very large extent. It could also be seen that the RMS value underestimated the peak values significantly.

To quantify the average peak amplitude, it was suggested to define the term Vibration Peak Magnitude (VPM) [7] (in Chapter 3.6.2 in Ref. [7], the former term VSL was used.) for a sampled signal a_n, with N data points by:

$$VPM(a_n) = \sqrt{\frac{\sum_{n=1}^{N} a_n^6}{\sum_{n=1}^{N} a_n^4}} \tag{1}$$

The VPM value is a weighted average of the peaks that obtains a representative peak acceleration in m/s^2. The VPM value for the impact wrench is indicated by the green dashed line in Figure 1.

2.2. Defining Shock Vibration

The suggestion was to define the term Vibration Shock Index (VSI) [7] (Chapter 3.6.2) as:

$$VSI = \frac{VPM}{RMS} \times \sqrt{2/3} \tag{2}$$

The VSI value is a dimensionless number that indicated to what extent a given signal consisted of shocks. It was defined as the ratio of the VPM and the RMS values, multiplied by the factor $\sqrt{2/3}$. With this definition, the VSI is defined as 1 for harmonic signals and increases when the shock content increases.

3. Results

Table 1 shows the values VSI, VPM, RMS and 1250 Hz lowpass filtered RMS to examples of machines with various types of vibration.

Table 1. Typical values from different machine types.

Machine	VSI	VPM [m/s^2]	RMS [m/s^2]	RMS Lowpass 1250 Hz [m/s^2]
Angle grinder	2.1	721	344	21
Impact wrench	11.4	7840	690	78
Nail gun	48.7	1520	31	16
Reciprocating saw	6.2	1170	188	50

4. Discussion

The VPM represents an effective estimate of the characteristic peak acceleration, while the VSI can be used for evaluating the degree of shock content. It is also of fundamental importance that if shock vibrations are to be evaluated, they are measured at frequencies high enough to cover at least a major part of the signal energy. Applying a low pass filter at 1250 Hz or ISO 5349-1 weighting vastly reduces the peak amplitude. As a compromise between the complexity of measurement equipment and covering the high-frequency content, it is suggested to measure at least up to 10 kHz and sampling at 50 kHz.

If there is a need of separating shock vibration from continuous, a suggestion is to define shock vibration as machines with a VSI exceeding the value of 8, which is believed to be in line with the general opinion. In general, a higher VSI indicates a higher importance of the shocks relative to the continuous content, though this needs to be further investigated as more data becomes available.

Enabling the possibility to quantify and define shock vibration will have a plurality of benefits, such as:

- Enhancing research on health effects, especially epidemiological studies;
- Enabling machine users to restrict the high-frequency content and develop demand specifications at procurement;
- Creating incentives for machine producers to reduce high-frequency vibration emission;
- Creating incentives for development of mitigation measures on existing tools;
- Promoting the development of personal protection equipment.

Author Contributions: Conceptualization, P.J. and H.L.; methodology, P.J. and H.L.; software, P.J. and S.L.G.; validation, H.L. and S.L.G.; writing—original draft preparation, H.L.; project administration, H.L.; funding acquisition, H.L. All authors have read and agreed to the published version of the manuscript.

Funding: AFA Insurance, #180109 and The Swedish Innovation board VINNOVA, #2020-04395.

Institutional Review Board Statement: Not applicable.

Informed Consent Statement: Not applicable.

Data Availability Statement: Not applicable.

Conflicts of Interest: The authors declare no conflict of interest.

References

1. Barregard, L.; Ehrenström, L.; Marcus, K. Hand-arm vibration syndrome in Swedish car mechanics. *Occup. Environ. Med.* **2003**, *60*, 287–294. [CrossRef] [PubMed]
2. Dandanell, R.; Engström, K. Vibration from riveting tools in the frequency range 6 Hz-10 MHz and Raynaud's phenomenon. *Scand. J. Work Environ. Health* **1986**, *12*, 338–342. [CrossRef] [PubMed]
3. Gerhardsson, L.; Ahlstrand, C.; Ersson, P.; Gustafsson, E. Vibration-induced injuries in workers exposed to transient and high frequency vibrations. *J. Occup. Med. Toxicol.* **2020**, *15*, 18. [CrossRef] [PubMed]
4. Lindell, H.; Lönnroth, I.; Ottertun, H. Transient Vibration from Impact Wrenches: Vibration Negative Effect on Blood Cells and Standards for Measurement. Presented at the Eighth International Conference on Hand-Arm Vibration, Umeå, Sweden, 9–12 June 1998.
5. Ando, H.; Nieminen, K.; Toppila, E.; Starck, J.; Ishitake, T. Effect of impulse vibration on red blood cells in vitro. *Scand. J. Work Environ. Health* **2005**, *31*, 286–290. [CrossRef] [PubMed]
6. Govinda Raju, S.; Rogness, O.; Persson, M.; Bain, J.; Riley, D. Vibration from a Riveting Hammer Causes Severe Nerve Damage in the Rat Tail Model. *Muscle Nerve* **2011**, *44*, 795–804. [CrossRef] [PubMed]
7. Johannisson, P.; Lindell, H. Definition and Quantification of Shock/Impact/Transient Vibrations. *arXiv* **2022**. [CrossRef]

 *proceedings*

Proceeding Paper

Daily Exposure Estimation from Field Measurements of Repetitive Shock Vibration [†]

Frédéric Maître [1] and Maël Amari [2],*

[1] Caisse Régionale d'Assurance Maladie d'Ile-de-France (CRAMIF), 75019 Paris, France; frederic.maitre@assurance-maladie.fr

[2] Electromagnetism, Vibration, Optics Laboratory, Institut National de Recherche et de Sécurité (INRS), 54519 Vandoeuvre-lès-Nancy, France

* Correspondence: mael.amari@inrs.fr; Tel.: +33-3-83-50-21-24

† Presented at the 15th International Conference on Hand-Arm Vibration, Nancy, France, 6–9 June 2023.

Abstract: Some hand-held power tools generate repeated shocks of high amplitudes. The daily vibration exposure $A(8)$ is rarely evaluated over the whole working day. The vibration total value a_{hv} emitted by a machine and the total daily duration of exposure T are usually estimated from a sample that may not be representative of the whole working task. Thus, the comparison of $A(8)$ with regulatory limit values is difficult. The daily number of shocks can sometimes be estimated from production features (i.e., the number of manufactured parts). This article presents an $A(8)$ evaluation method adapted to such situations. It allows for a better evaluation of the risk for non-stationary vibrations.

Keywords: hand-arm vibration; multiple shocks; daily vibration exposure; vibration total value; total daily duration of exposure; evaluation method

1. Introduction

The standard method for assessing exposure to hand-arm vibration is defined by ISO 5349-1 [1]. The daily vibration exposure $A(8)$ of an operator must be derived from the vibration total value a_{hv} emitted by the machine and the total daily duration of exposure T. In the field, the evaluation of $A(8)$ is often performed on a limited time period in order not to disturb the company's production.

Some hand-held tools generate shocks and transient vibrations of high amplitudes [2]. For these non-stationary signals, the vibration total value a_{hv} and the total daily exposure time T may not be representative of the worker's real exposure: irregular working task, signal splitting by the experimenter, and limited information on daily activities. The estimation of the daily vibration exposure $A(8)$ and its comparison with the regulatory limit values are therefore difficult [3].

The aim of this article is to present a method for the estimation of the daily vibration exposure $A(8)$ adapted to vibrations composed of repeated shocks. It only requires the calculation of the vibration dose received by the operator during the intervention and the estimation of the total number of daily shocks. Its mathematical formulation is presented. It is then illustrated on a real case of field measurement.

2. Materials and Methods

2.1. Daily Vibration Exposure

The daily vibration exposure $A(8)$ is defined as:

$$A(8) = a_{hv} \cdot \sqrt{\frac{T}{T_0}} \left(\mathrm{m \cdot s^{-2}} \right), \tag{1}$$

a_{hv} (m·s^{-2}) is the vibration total value emitted by the machine. T (s) is the total daily duration of exposure of the operator. $T_0 = 8 \times 3600$ (s) is the reference duration for a working day.

The vibration total value emitted by the machine a_{hv} is defined as:

$$a_{hv} = \sqrt{a_{hw_x}{}^2 + a_{hw_y}{}^2 + a_{hw_z}{}^2}\left(\text{m·s}^{-2}\right),\tag{2}$$

$a_{hw_{x,y,z}}$ (m·s^{-2}) is the root-mean-square value of the W_h weighted accelerations, measured in the directions x, y, and z.

The root-mean-square value of the W_h weighted accelerations $a_{hw_{x,y,z}}$ is defined as:

$$a_{hw_{x,y,z}} = \sqrt{\frac{1}{N} \cdot \sum_{n=1}^{N} \left|a_{hw_{x,y,z}}[n]\right|^2}\left(\text{m·s}^{-2}\right),\tag{3}$$

$a_{hw_{x,y,z}}[n]$ (m·s^{-2}) is the nth sample value of the W_h weighted acceleration. $N = T \cdot f_s$ is the total number of samples. f_s (Hz) is the sampling frequency of the signal.

2.2. Estimation from Field Measurements

$a_{hv_{sample}}$, T_{sample}, and $A(8)_{sample}$ are the quantities related to the measurements made over a portion of the working day (measured work phase). In particular, T_{sample} is the sample duration over which $a_{hv_{sample}}$ is calculated.

$a_{hv_{estimate}}$, $T_{estimate}$, and $A(8)_{estimate}$ are the estimates of the unknown quantities a_{hv}, T, and $A(8)$ relative to the whole working day.

If the measured work phase is considered to be representative of the operator's daily task, then $A(8)_{estimate}$ may be derived from Equation (1) as:

$$A(8)_{estimate} = a_{hv_{sample}} \cdot \sqrt{\frac{T_{estimate}}{T_0}}\quad\left(\text{m·s}^{-2}\right),\tag{4}$$

Depending on the nature of the working task, it is sometimes possible to know the numbers of shock repetitions during the measured work phase R_{sample}, and during complete working task $R_{estimate}$. $A(8)_{estimate}$ is then derived from Equation (1) as:

$$A(8)_{estimate} = \sqrt{R_{estimate}} \cdot \overline{A_r(8)}_{sample}\left(\text{m·s}^{-2}\right),\tag{5}$$

$\overline{A_r(8)}_{sample}$ (m·s^{-2}) represents the average vibration dose received per shock r of the sample. It is equal to the dose that would be received for each of the R_{sample} shocks of the sample if they were all identical:

$$\overline{A_r(8)}_{sample} = \frac{A(8)_{sample}}{\sqrt{R_{sample}}}\quad\left(\text{m·s}^{-2}\right),\tag{6}$$

2.3. Application to Vibrations Emitted by Firearms

Vibrations emitted by a Zastava M70 AB2 automatic assault rifle were measured (Figure 1). Five 7.62×39 mm M43 rounds were fired in 35.5 s during the measurements. The ballistics expert estimates that this weapon is used for 1.5 h per day. Another piece of information is that approximately 300 rounds are fired per day both in semi-automatic and full-automatic mode. A piezoelectric accelerometer was glued to the rifle (356B20-PCB Piezotronics®). Acceleration was recorded with a R2DB front-end (DEWESOFT®).

Figure 1. Measurement of vibrations emitted by a Zastava M70 AB2 automatic assault rifle. Top: riffle overview; bottom left: accelerometer positioning; bottom right: 7.62 × 39 mm M43 rounds.

3. Results

3.1. Field Measurements

Figure 2 presents the weighted acceleration $a_{hw_y}[n]$ of the measured work phase. Only one axis is shown for readability reasons. An extract of the operator's daily working task is also presented in order to illustrate the differences that may exist with the sample. This part remains unknown during field measurements. Signals characteristics are presented in Table 1.

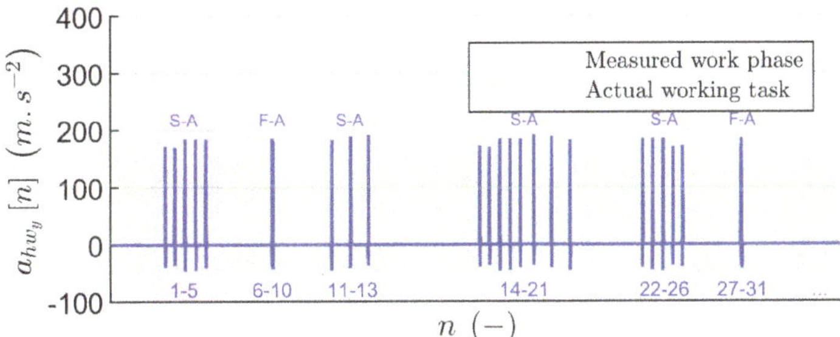

Figure 2. Extract from the field measurement: weighted acceleration $a_{hw_y}[n]$ and number of shots fired (S-A: semi-automatic, F-A: full-automatic).

Table 1. Vibration exposure characteristics: sample (measured work phase) and daily working task.

Measured Work Phase					Actual Working Task [1]				
$a_{hv_{sample}}$ (m·s^{-2})	T_{sample} (s)	$A(8)_{sample}$ (m·s^{-2})	R_{sample} $(-)$	$\overline{A_r(8)}_{sample}$ (m·s^{-2})	a_{hv} (m·s^{-2})	T (s)	$A(8)$ (m·s^{-2})	R $(-)$	$\overline{A_r(8)}$ (m·s^{-2})
2.6	35.5	0.090	5	0.040	2.1	3500.0	0.7	310	0.041

[1] Unknown.

3.2. Daily Vibration Exposure Estimations

Table 2 shows $A(8)_{estimate}$ values calculated from the sample depending on the available information.

Table 2. Daily vibration exposure estimations from sample (measured work phase) and available information.

	$T_{estimate}$ [1]			$R_{estimate}$ [1]		
$a_{hv\,sample}$ (m·s^{-2})	$T_{estimate}$ (s)	$A(8)_{estimate}$ (m·s^{-2})	$R_{estimate}$ $(-)$	R_{sample} $(-)$	$A(8)_{sample}$ (m·s^{-2})	$A(8)_{estimate}$ (m·s^{-2})
2.6	5400	1.1	300	5	0.090	0.7

[1] Known from task information.

4. Discussion

For non-stationary vibrations composed only of one or more successive shocks, Equations (1)–(3) show that a_{hv} varies with the number of points N of a signal. This is not the case for $A(8)$.

When only the total daily duration of exposure $T_{estimate}$ can be determined from the daily task information, Equation (4) shows that $A(8)_{estimate}$ depends on $a_{hv\,sample}$. $A(8)_{estimate}$ is then particularly sensitive to the sampling conditions: the machine emission, and the regularity of the firing rate throughout the day.

When the numbers of shocks of the sample R_{sample} and of the whole day $R_{estimate}$ can be determined from task information, Equations (5) and (6) show that $A(8)_{estimate}$ depends only on the dose $A(8)_{sample}$. $A(8)_{estimate}$ is then insensitive to the consistency of the firing pace outside of the sample.

In practice, the field-measurement conditions are not always controlled. $a_{hv\,sample}$ varies during the day depending on the company's production (Figure 2). The assumption that the sample is representative is not always satisfied (Table 1). $T_{estimate}$ is also difficult to determine as it represents the actual duration of exposure to vibration. It can be very different from the duration of use of the machine. The estimate of daily vibration exposure $A(8)_{estimate}$ is then strongly biased (Table 2).

When possible, the estimation of the total number of shocks $R_{estimate}$ is easier to perform and often more accurate than $T_{estimate}$. The average vibration dose received per shock during sampling $\overline{A_r(8)}_{sample}$ is often representative of the whole set of shocks (Table 1). $A(8)_{estimate}$ is then closer to the actual daily exposure.

5. Conclusions

The estimation method for the daily exposure to vibration $A(8)$, which takes into account the number of shock repetitions, is often more reliable than the usual one. It also facilitates the implementation of technical prevention solutions by linking the operator's exposure to production rather than to his work pace. It should be preferred for assessing the risk created by repeated shocks at the workplace.

Author Contributions: Conceptualization, F.M. and M.A.; methodology, F.M.; software, F.M.; validation, F.M. and M.A.; formal analysis, F.M.; investigation, F.M.; resources, F.M.; data curation, F.M.; writing—original draft preparation, F.M.; writing—review and editing, F.M.; visualization, F.M.; supervision, F.M.; project administration, F.M.; funding acquisition, F.M. All authors have read and agreed to the published version of the manuscript.

Funding: This research received no external funding.

Institutional Review Board Statement: Not applicable.

Informed Consent Statement: Not applicable.

Data Availability Statement: Data not available due to legal/commercial restrictions.

Conflicts of Interest: The authors declare no conflict of interest.

References

1. *ISO 5349-1:2001*; Mechanical Vibration—Measurement and Evaluation of Human Exposure to Hand-Transmitted Vibration—Part 1: General Requirements. International Organization for Standardization: Geneva, Switzerland, 2001.
2. Pitts, P.; Kaulbars, U.; Lindell, H.; Grétarsson, S.L.; Machens, M.; Brammer, A.J.; Yu, G.-Q.; Schenk, T.; Haas, F. *Hand-Arm Vibration: Exposure to Isolated and Repeated Shock Vibrations—Review of the International Expert Workshop 2015 in Beijing*; IFA Report 5/2017e; IFA: Bonn, Germany, 2015; ISBN 978-3-86423-198-8.
3. Directive 2002/44/EC of the European Parliament and of the Council of 25 June 2002 on the Minimum Health and Safety Requirements Regarding the Exposure of Workers to the Risks Arising from Physical Agents (vibration). Available online: https://eur-lex.europa.eu/resource.html?uri=cellar:546a09c0-3ad1-4c07-bcd5-9c3dae6b1668.0004.02/DOC_1&format=PDF (accessed on 1 December 2022).

Proceeding Paper

The Vibration Characteristics of Ultrasonic-Activated Straightening and Forming Machines †

Didier Aoustin

Carsat Bretagne, Département des Risques Professionnels, Centre Interrégional de Mesures Physiques de l'Ouest (CIMPO), 35000 Rennes, France; didier.aoustin@carsat-bretagne.fr; Tel.: +33-299267312

† Presented at the 15th International Conference on Hand-Arm Vibration, Nancy, France, 6–9 June 2023.

Abstract: The straightening and forming of metal parts are typical and usual tasks within shipbuilding industries. They may expose boilermakers to high levels of vibration, leading to potential health risks (musculoskeletal disorders and vascular syndromes). This study consisted of helping a company evaluate their hammering machines that were activated by ultrasound, comparing them to conventional impact tools. Vibration measurements were therefore carried out in the field. These were followed by an analysis of the signals according to the regulations and additional assessment tools.

Keywords: hand arm vibration; straightening; hammering activated by ultrasound

Citation: Aoustin, D. The Vibration Characteristics of Ultrasonic-Activated Straightening and Forming Machines. *Proceedings* **2023**, *86*, 5. https://doi.org/10.3390/proceedings2023086005

Academic Editors: Christophe Noël and Jacques Chatillo

Published: 7 April 2023

1. Introduction

Context: To achieve the final shape of a component or a metal part, it is often necessary to carry out straightening or forming operations. These activities are common in aeronautical and naval construction. They are conventionally performed by hammering processes that use different tools: (i) a sledgehammer and flattener; (ii) a trolley-mounted pneumatic chipping hammer; or (iii) a pneumatic riveting hammer.

These operations can expose personnel to high vibration and sound levels. One technology that is present in the market is the ultrasound-activated hammer [1]. A high-frequency generator supplies a piezoelectric transmitter, generating mechanical waves. These are amplified to the sonotrode and this energy is then transmitted to a media, which is made up of needle-type impactors for the straightening–forming.

Purpose: To support a shipbuilding company in its prevention approach during the qualification tests of a hammering process activated by ultrasound, we characterised the vibration emissions of this technology and compared them to conventional tools.

2. Materials and Method

2.1. Tools Tested and Test Conditions

We first tested the manual hammering technique with a flattener (Figure 1).

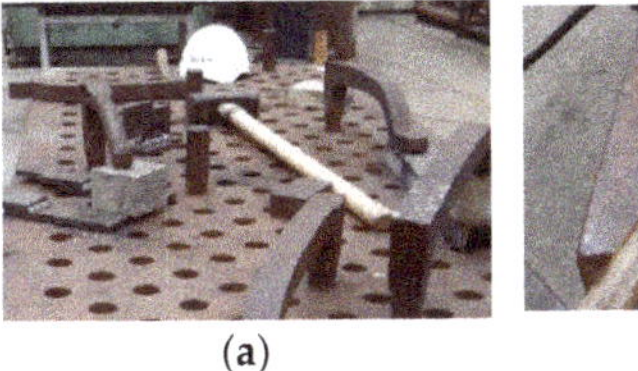
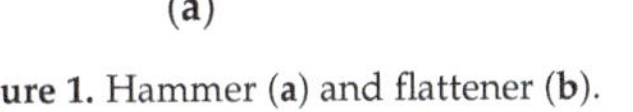

Figure 1. Hammer (**a**) and flattener (**b**).

Then, a riveting hammer and a chipping hammer on a trolley were characterised (Figure 2).

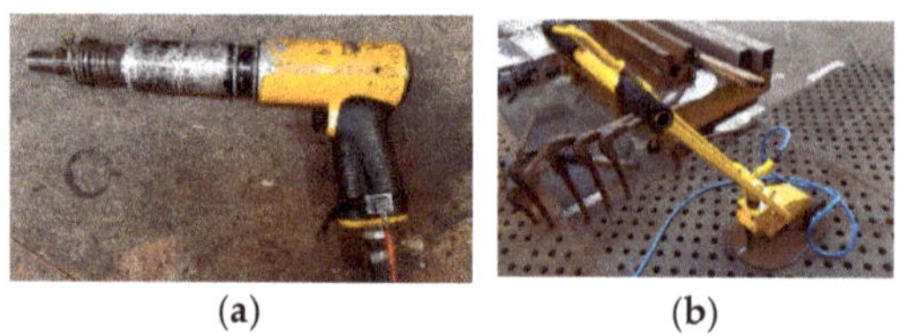

(a)   (b)

Figure 2. Atlas Copco RRH 12 (**a**), and (**b**) RRC 75 B-01.

Finally, we characterised two hammers that were activated by ultrasound (Figure 3).

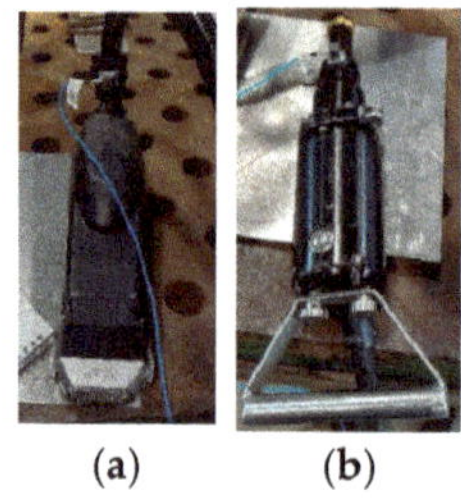

(a)   (b)

Figure 3. SONATS Stress Voyager PR10 mouthpiece ER18-03 (**a**), and SONATS Nomad (**b**).

The parts that were used were made up of metal elements of steel or aluminium, clamped on a test marble in the boiler-making workshop: (i) steel and aluminium sheets; and (ii) mechanically welded steel parts.

2.2. Metrology

The measurements of the hand-transmitted vibrations were carried out in accordance with the NF EN ISO 5349-1 and NF EN ISO 5349-2 [2,3] standards, using a triaxial accelerometer that was rigidly fastened to the machine handle, as close as possible to the palm of the hand.

The signal conditioning of our triaxial accelerometer (model 356B20, PCB Piezotronics) and the vibration data acquisition were carried out using our acquisition front-end (Scadas XS, Siemens), with a sampling rate of 2560 Hz.

2.3. Data Processing

2.3.1. Vibration Frequency Weighting

The signals that were measured on the three axes are weighted in frequency, according to the weighting factor W_h, as defined by the NF EN ISO 5349-1 standard [2]. The corresponding weighted accelerations are expressed as a_{hwx}, a_{hwy}, and a_{hwz}.

The documentation booklet FD ISO/TR 18570 [4] defines an additional method for measuring the vibrations that are transmitted to the hands. It aims at improving the assessment of the risks of vascular disorders (vibration white finger, namely Raynaud syndrome).

This booklet defines, in particular, an additional frequency weighting, W_p (Figure 4), for the three axes, x, y, and z. The main difference between this weighting and the weighting W_h, as defined by the NF EN ISO 5349-1 standard [1], lies in its much greater consideration of the higher frequencies.

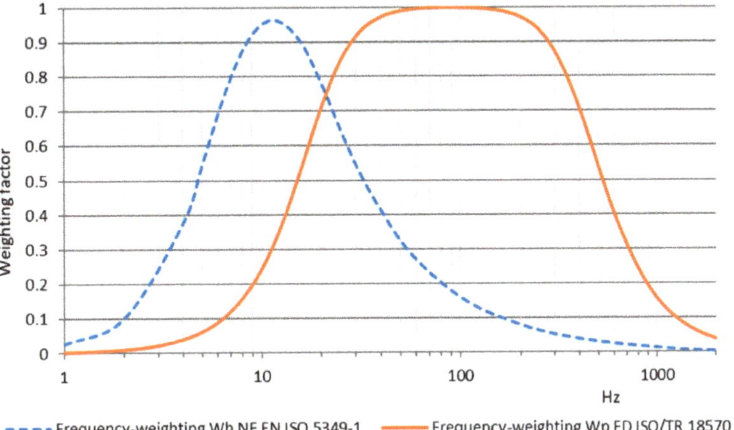

Figure 4. Weighting curve Wp to assess the risk of vascular disorders.

2.3.2. Determination of Vibration Emission Levels

The level of the vibration emission, a_{hv} (the vibration total value), is calculated according to the following expression (1):

$$a_{hv} = \sqrt{a_{hwx}{}^2 + a_{hwy}{}^2 + a_{hwz}{}^2} \tag{1}$$

As the a_{hv} level is, the level of the vibration emission a_{pv} (the vibration total value) is a function of the frequency-weighted accelerations that are measured on the three axes (a_{px}, a_{py}, and a_{pz}) and is calculated by the following expression (2):

$$a_{pv} = \sqrt{a_{px}{}^2 + a_{py}{}^2 + a_{pz}{}^2} \tag{2}$$

2.3.3. Spectral Analyses of Temporal Signals

Thanks to our SiemensScadas XS system and its TestXpress software, from the time signals, we analysed the vibration spectra of the hammering that was activated by ultrasound.

3. Results

The qualification tests of the hammering tools activated by ultrasound, in comparison to the conventional straightening tools, did not provide satisfaction to the company.

3.1. Vibration Levels of Conventional Processes

We calculated the vibration levels of the standard tools that were used by this shipbuilding company for the straightening and forming of parts (Table 1):

Table 1. Vibration levels of conventional tools.

Measured Configuration	Vibration Level (m/s^2)
	a_{hv}
Riveting hammer Atlas Copco RRH12	6.3
Chipping hammer Atlas Copco RRC 75B-01	3.3
Flattener (7 shots—duration: 15 s)	12.9
Masse (5 shots—duration: 9 s)	21.9

3.2. Vibration Levels of Ultrasound hammer

We calculated the vibration levels of the ultrasonically activated hammering tools that were tested for the straightening and forming of parts (Table 2).

Table 2. Vibration levels of ultrasonically activated tools.

Measured Configuration	Vibration Level (m/s^2)	
	a_{hv}	a_{pv}
Head PR10 aluminium sheet	4.6	26.0
Head PR10, steel piece	4.8	28.1
Head PR17, steel piece, main handle	1.0	9.5
Head PR17, aluminium sheet, main handle	1.2	13.1
Head PR17, aluminium sheet, accelerometer on head	1.5	14.4
Head PR17, steel piece, accelerometer on head	1.3	14.8
Head PR13, aluminium sheet	2.4	10.4
Head PR13, steel sheet	1.7	13.1
Nomad, aluminium sheet, rear handle	4.7	14.6
Nomad, steel sheet, rear handle	4.9	23.1
Nomad, aluminium sheet, accelerometer on head	3.7	10.2
Nomad, steel sheet, accelerometer on head	4.0	18.7

4. Discussion

Compared to conventional tools, ultrasound-activated hammering tools generally reduce vibration levels. There are, however, strong variations depending on the type of head and mouthpiece that are used for the same piece being hammered, with factors ranging from 1 to 4. This leads to limit durations of use of just over 2 h for reaching the regulatory action threshold value set at 2.5 m/s^2. Our readings show significant spectral content in high frequencies (50 to 500 Hz). The use of the FD ISO/TR 18570 documentation booklet allowed us to estimate the vibration levels and their vascular risks. Without a limit value, however, it is difficult to assess the level of risk.

5. Conclusions

Synthesis: These new machines reduce risks when straightening and forming metal parts. However, we have noted that there are health risks depending on the type of tools that are used. These tests did not provide satisfaction to the shipbuilding company and the traditional processes were therefore retained.

Prospects: To characterise these new tools under operational conditions, giving satisfaction to the user, we have planned to carry out new tests with the manufacturer of these processes, in an establishment of one of their customers.

Funding: This research received no external funding.

Institutional Review Board Statement: Not applicable.

Informed Consent Statement: Not applicable.

Data Availability Statement: Data not available due to legal/commercial restrictions.

Conflicts of Interest: The author declares no conflict of interest.

References

1. STRESSONIC® Peening Process. Available online: https://sonats-et.com/en/peening-process/ (accessed on 15 February 2023.).
2. *NF EN ISO 5349-1:2002*; Mechanical Vibration—Measurement and Evaluation of Human Exposure to Hand-Transmitted Vibration—Part 1: General requirements. International Organization for Standardization: Geneva, Switzerland, 2002.

3.   *NF EN ISO 5349-2:2001*; Mechanical Vibration—Measurement and Evaluation of Human Exposure to Hand-Transmitted Vibration—Part 2: Practical guidance for measurement at the workplace. International Organization for Standardization: Geneva, Switzerland, 2001.
4.   *FD ISO/TR 18570:2017*; Mechanical Vibration—Measurement and Evaluation of Human Exposure to Hand Transmitted Vibration—Supplementary Method for Assessing Risk of Vascular Disorders. International Organization for Standardization: Geneva, Switzerland, 2017.

Proceeding Paper

Hand-Arm Vibration Exposure Assessment for a Case-Control Study among German Workers [†]

Christian Freitag [1], Yi Sun [1], Frank Bochmann [1], Winfried Eckert [2], Benjamin Ernst [1], Uwe Kaulbars [1], Uwe Nigmann [3], Christina Samel [1], Christian van den Berg [4] and Nastaran Raffler [1,*]

[1] Institute for Occupational Safety and Health of the German Social Accident Insurance (IFA), 53757 Sankt Augustin, Germany; christian.freitag@dguv.de (C.F.); yi.sun@dguv.de (Y.S.); frank.bochmann@dguv.de (F.B.); benjamin.ernst@dguv.de (B.E.); uwe.kaulbars@dguv.de (U.K.); christina.samel@dguv.de (C.S.)

[2] German Social Accident Insurance Institution for the Building Trade, 10715 Berlin, Germany; winfried.eckert@bgbau.de

[3] German Social Accident Insurance Institution for the Woodworking and Metalworking Industries, 55124 Mainz, Germany; uwe.nigmann@bghm.de

[4] German Social Accident Insurance Institution for the Raw Materials and Chemical Industry, 69115 Heidelberg, Germany; christian.vandenberg@bgrci.de

[*] Correspondence: nastaran.raffler@dguv.de

[†] Presented at the 15th International Conference on Hand-Arm Vibration, Nancy, France, 6–9 June 2023.

Abstract: In order to analyse the exposure-response relationship between hand–arm-vibration exposure and the risk of musculoskeletal disorders (MSDs) of upper extremities in an epidemiological case-control study, a database was established to provide technical characteristics such as vibration exposure data and its frequency component measured at workplaces. This hand-arm vibration database consists of over 730 technical tools and devices, whereas 422 devices were used for exposure assessment in the epidemiological case-control study. The devices used were divided into a total of 13 device groups: hammers, grinding machines, compactors, screwdrivers and saws were the most frequently used devices.

Keywords: hand-arm vibration exposure; database; frequency components

Citation: Freitag, C.; Sun, Y.; Bochmann, F.; Eckert, W.; Ernst, B.; Kaulbars, U.; Nigmann, U.; Samel, C.; van den Berg, C.; Raffler, N. Hand-Arm Vibration Exposure Assessment for a Case-Control Study among German Workers. *Proceedings* **2023**, *86*, 41. https://doi.org/10.3390/proceedings2023086041

Academic Editors: Christophe Noël and Jacques Chatillon

Published: 17 May 2023

1. Introduction

According to expert estimates, there are currently around 1.5 to 2 million employees in Germany who are exposed to hand-arm vibration [1]. In Germany, there are two different diseases caused by hand-arm vibration exposure. The most extensively studied forms of hand-arm vibration syndrome are vascular and neurological disorders [2]. Recent studies and reviews also indicate an elevated risk of musculoskeletal symptoms and osteoarthritis among vibration-exposed workers [2–5]. However, so far, an exposure-response relationship between hand-arm vibration exposure and MSDs of upper extremities has been poorly established.

In order to evaluate the exposure–response relationship between hand–arm vibration exposure and the risk of MSDs, an epidemiological case-control study was carried out among workers in the construction, mining, metal and wood working industries in Germany. Within this project, a database was to establish providing physical characteristics of the investigated tools. The database was designed to provide information about the vibration exposure but also working conditions and workpieces, which can be used for retrospective analysis.

2. Material and Methods

To quantify individual vibration exposures, a database of industrial hygiene measurements of over 700 power tools at workplaces was established. This database allows for detailed quantification of vibration exposures over time.

According to international standard ISO 5349-1:2001 [6], vibration generated by the technical power tools was measured in three orthogonal directions (x, y, and z). Vibration values are expressed as accelerations a_{hwx}, a_{hwy}, and a_{hwz} in the three measuring directions. The vibration total value (a_{hv}) was determined as the root-sum-of-squares of the three component values (a_{hwx}, a_{hwy} and a_{hwz}):

$$a_{hv} = \sqrt{a_{hwx}^2 + a_{hwy}^2 + a_{hwz}^2} \tag{1}$$

In addition to the vibration total value (a_{hv}), acceleration (a_{hw}) in the direction along the forearm (direction z) was also considered in the exposure–response analyses of this study.

Based on the guideline VDI 2057, part 2 [7], the frequency components of the vibration exposure for the forearm direction have been detected for the following frequency ranges: 4 to 20 Hz, 4 to 31.5 Hz, 4 to 50 Hz, 4 to 80 Hz, and 4 to 100 Hz. These frequency components could be a distinguishing feature for determining the low- and high-frequency devices.

3. Results and Conclusions

The hand-arm vibration exposure for the forearm direction and the vibration total value with their frequency components of 422 devices are given in Figure 1.

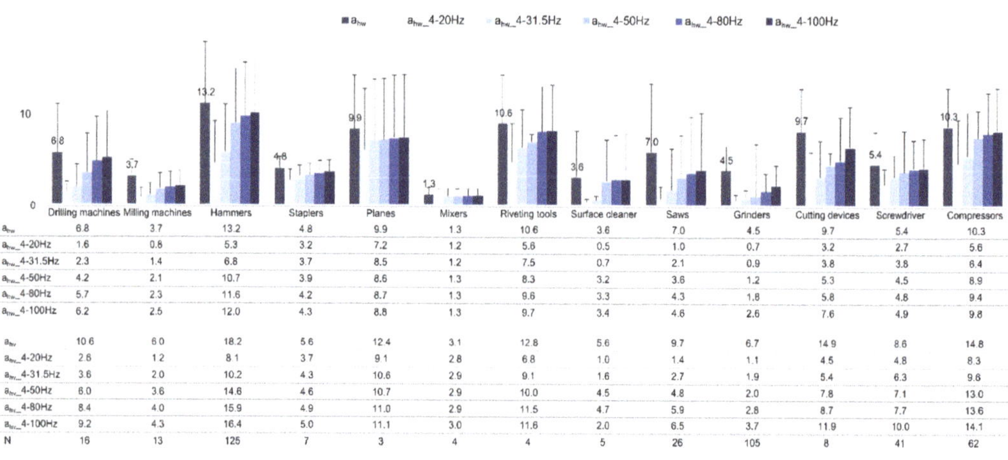

	Drilling machines	Milling machines	Hammers	Staplers	Planes	Mixers	Riveting tools	Surface cleaner	Saws	Grinders	Cutting devices	Screwdriver	Compressors
a_{hw}	6.8	3.7	13.2	4.8	9.9	1.3	10.6	3.6	7.0	4.5	9.7	5.4	10.3
a_{hw}_4-20Hz	1.6	0.8	5.3	3.2	7.2	1.2	5.6	0.5	1.0	0.7	3.2	2.7	5.6
a_{hw}_4-31.5Hz	2.3	1.4	6.8	3.7	8.5	1.2	7.5	0.7	2.1	0.9	3.8	3.8	6.4
a_{hw}_4-50Hz	4.2	2.1	10.7	3.9	8.6	1.3	8.3	3.2	3.6	1.2	5.3	4.5	8.9
a_{hw}_4-80Hz	5.7	2.3	11.6	4.2	8.7	1.3	9.6	3.3	4.3	1.8	5.8	4.8	9.4
a_{hw}_4-100Hz	6.2	2.5	12.0	4.3	8.8	1.3	9.7	3.4	4.6	2.6	7.6	4.9	9.8
a_{hv}	10.6	6.0	18.2	5.6	12.4	3.1	12.8	5.6	9.7	6.7	14.9	8.6	14.8
a_{hv}_4-20Hz	2.6	1.2	8.1	3.7	9.1	2.8	6.8	1.0	1.4	1.1	4.5	4.8	8.3
a_{hv}_4-31.5Hz	3.6	2.0	10.2	4.3	10.6	2.9	9.1	1.6	2.7	1.9	5.4	6.3	9.6
a_{hv}_4-50Hz	6.0	3.6	14.6	4.6	10.7	2.9	10.0	4.5	4.8	2.0	7.8	7.1	13.0
a_{hv}_4-80Hz	8.4	4.0	15.9	4.9	11.0	2.9	11.5	4.7	5.9	2.8	8.7	7.7	13.6
a_{hv}_4-100Hz	9.2	4.3	16.4	5.0	11.1	3.0	11.6	2.0	6.5	3.7	11.9	10.0	14.1
N	16	13	125	7	3	4	4	5	26	105	8	41	62

Figure 1. Hand-arm vibration exposure assessment for the forearm direction, vibration total value and their frequency components.

Hammers followed by riveting tools, compressors, planes and cutting devices are the most vibration generating devices (13.2 ms⁻², 10.6 ms⁻², 10.3 ms⁻², 9.9 ms⁻² and 9.7 ms⁻² respectively for forearm direction). The lowest vibration exposure in this project was assessed for the mixers with 1.3 ms⁻² in the forearm direction.

Regarding the frequency components, staples, planes and mixers did not show any notable differences between the frequency ranges. On the contrary, hammers, surface cleaner, saws, cutting devices and compressors showed big differences within the frequency ranges, especially regarding the range between 4 and 50 Hz.

Author Contributions: Conceptualization, methodology: C.F., Y.S., F.B., W.E., B.E., U.K., U.N., C.S., C.v.d.B. and N.R. Software, validation, formal analysis, investigation, resources, data curation, writing—original draft preparation, writing—review and editing, visualization, supervision, project administration: N.R., U.K. and C.F. All authors have read and agreed to the published version of the manuscript.

Funding: This research was funded by the German Social Accident Insurance, grant number FP-297.

Institutional Review Board Statement: Not applicable.

Informed Consent Statement: Informed consent was obtained from all subjects involved in the study.

Data Availability Statement: The data that support the findings of this study are available from the corresponding author, [N.R.], upon reasonable request.

Conflicts of Interest: The authors declare no conflict of interest.

References

1. Mohr, D. *Eine Einfache Methode zur Beurteilung Stoßhaltiger Ganzkörper-Schwingungen*; VDI Tagung Humanschwingungen; VDI Verlag: Duesseldorf, Germany, 2004.
2. House, R.; Wills, M.; Liss, G.; Switzer-McIntyre, S.; Manno, M.; Lander, L. Upper extremity disability in workers with hand-arm vibration syndrome. *Occup. Med.* **2009**, *59*, 167–173. [CrossRef] [PubMed]
3. Hagberg, M. Clinical assessment of musculoskeletal disorders in workers exposed to hand-arm vibration. *Int. Arch. Occup. Environ. Health* **2002**, *75*, 97–105. [CrossRef] [PubMed]
4. Bovenzi, M. Health risks from occupational exposures to mechanical vibration. *Med. Lav.* **2006**, *97*, 535–541. [PubMed]
5. Charles, L.E.; Ma, C.C.; Burchfiel, C.M.; Dong, R.G. Vibration and Ergonomic Exposures Associated With Musculoskeletal Disorders of the Shoulder and Neck. *Saf. Health Work* **2018**, *9*, 125–132. [CrossRef] [PubMed]
6. *ISO 5349-1:2001*; Mechanical Vibration—Measurement and Evaluation of Human Exposure to Hand-Transmitted Vibration—Part 1: General Requirements. International Organization for Standardization: Geneva, Switzerland, 2001.
7. *VDI 2057 BLATT 2*; Human Exposure to Mechanical Vibration—Hand-Arm Vibration. VDI Verlag: Duesseldorf, Germany, 2016.

Proceeding Paper

Necessity and Considerations for On-Body Vibration Measurement Equipment †

Setsuo Maeda [1],* , Ying Ye [2] and Shuxiang Gao [2]

1   School of Science and Technology, Nottingham Trent University, Clifton Lane, Nottingham NG11 8NS, UK
2   Human Factors Research Unit, Institute of Sound and Vibration Research, University of Southampton, Southampton SO17 1BJ, UK; y.ye@soton.ac.uk (Y.Y.); s.gao@soton.ac.uk (S.G.)
*   Correspondence: setsuo.maeda@ntu.ac.uk; Tel.: +44-115-941-8418
†   Presented at the 15th International Conference on Hand-Arm Vibration, Nancy, France, 6–9 June 2023.

Abstract: The palmar surface (on tool surface) has been defined in ISO 5349-1 as a value of the amount of vibration transmitting to the hand and arm from on-body vibration magnitude. They showed the concept of on-body vibration measurement based on the relationship between the temporary threshold shift (TTS) of the vibrotactile perception threshold (VPT) and the on-body vibration measurement values. However, they did not show that the effectiveness of ISO 5349-1 Annex D for various factors transmitting to the hand was unknown. Therefore, the purpose of this paper is to clarify the new considerations of on-body vibration measurement equipment and to demonstrate the necessity of on-body measurement equipment.

Keywords: hand-transmitted vibration; on-body vibration; vibration measurement equipment; vibrotactile perception

1. Introduction

A guideline for the measurement of hand-transmitted vibration was released internationally in ISO 5349-1 in 2001 [1], concerning the problem of hand–arm vibration exposure. At present, employers seek to follow the guidance within the ISO 5349-1 standard for preventing HAVS. Within clause 4.3 of this standard, it is stated that the acceleration measured at the surface of the vibrating tool in contact with the hand is used as the primary measurement to characterize the vibration exposure. Therefore, ISO 5349-1 assumes that the hand-transmitted vibration exposure magnitude is measured on the tool handle, although the hand-transmitted vibration is affected by many factors, as listed in Annex D of ISO 5349-1 standard. For many years, the factors outlined within Annex D of ISO 5349-1 have not been adequately accounted for when assessing hand-transmitted vibration exposure for the purposes of the prevention of hand–arm vibration syndrome (HAVS) in real work environments [2]. A desire by employers to adhere strictly to the ISO 5349-1 standard may be contributing to inaccurate dose assessments and inferior outcomes for the worker [3]. Although researchers have studied the effects of the vibration magnitude, their results cannot apply directly to evaluating and assessing the risk from hand-transmitted vibrations in a real work site. Since conducting a long-term study on a work site would present difficulties, as using a method such as on-tool measuring equipment would disturb work on-site, there is a need to develop a discreet new vibration exposure meter in order to understand the level of exposure of an individual at an actual work site. Based on these results, the vibration exposure dose transmitted to the worker's hand–arm system can be measured accurately, and consequently the development of vibration exposure equipment capable of vibration tool work is increasingly sought after. There are numerous factors which influence the vibration exposure, such as tool construction, tool condition, attachments used, attachment condition, material of the workpiece, direction of operation, operator posture, feed force, and grip force. Even if the same work was being conducted, the individual workers' vibration exposure level would also be highly dependent on their degree of skill, so the current risk assessment methods cannot take every influencing factor into consideration. Currently, the question of how to use the obtained vibration exposure data to predict workers' risk of hand transmitted vibration is the biggest problem.

Therefore, in this paper, the concept and necessity of equipment that predicts both on-tool vibration values that incorporate ISO 5349-1 Annex D from ISO 5349-1, and on-body vibration measurements, have been demonstrated.

2. Measurement on the Vibrating Surface

Measurements of hand–arm vibration in accordance with ISO 5349-1 and ISO 5349-2 require measurement on the interface between the vibrating surface and the hand. This usually requires a transducer set to be fixed to the gripping zone of a machine, with signal cables running to a measurement system positioned away from both the machine and the machine operator. Measurements on the vibrating surface using instrumentation that complies with ISO 8041-1 [4] or ISO 8041-2 [5] are required for compliance with ISO 5349-1. However, these measurements can be complex and require technical knowledge, skill, and experience to achieve reliable results. Full compliance with ISO 5349-1 may not be required if the purpose is to understand and control vibration exposure, or for research applications. In workplaces, workers are using tools in many ways and with different work postures. After prolonged tool usage, many workers suffer from hand-arm vibration syndromes, e.g., vascular diseases, neurological diseases, and musculoskeletal diseases.

Many researchers have worked on clarifying the relationship between dose and human responses to vibration, such as diseases, for many years. The vibration measurements were made in the field during real operating conditions performed by workers. The vibration was measured in three orthogonal directions according to the international standard ISO 5349-1 procedure (ISO 2001) on the tool handle. The vibration magnitudes were expressed as root–mean–square (r.m.s.) acceleration, and were frequency-weighted using frequency weighting W_h in accordance with ISO 5349-1 [1]. The root-sum-of-squares (vibration total value) of the frequency-weighted acceleration values a_{hv} for the x-, y-, and z-axes were calculated as shown in Equation (1).

$$a_{hv} = \left(a_{hwx}^2 + a_{hwy}^2 + a_{hwz}^2 \right)^{\frac{1}{2}} \tag{1}$$

3. Necessity and Considerations of On-Body Vibration Measurement Equipment

In dealing with some of these aforementioned issues, on-body vibration measurement equipment presents some attractive characteristics.

- Small size, light weight, and possibility of attachment to vibration measurement equipment.
- Solid structure that does not damage workability.
- Able to measure the vibration acceleration magnitude when added to a hand palm surface (on tool handle).
- Able to indicate a measured variable or to indicate an exposure point.
- Able to show a warning when exceeding the EAV (exposure action value) or ELV (exposure limit value).

The response of the vibration measured on the wrist is an alternative method proposed to evaluate the vibration exposure [3]. This estimated value gives a higher correlation with that of the palm surface, and from this it is judged to be a target value for evaluation, which is the reason the wrist was considered at a work site for vibration measurement.

It is extremely difficult to determine these effects of the vibration exposure magnitude on the human body during work. Annex D of ISO 5349-1 identifies several factors that impact the hand-transmitted vibration magnitude. The proposed consideration of this study, to account for factors affecting the transfer of the vibration magnitude from the tool handle to the on-body measurement equipment (on wrist), is to estimate the tool vibration magnitude by using Equation (2).

$$Estimated\ on\ tool\ vibration\ magnitude = fw * TR\left(\frac{A_{\text{handle}}}{A_{\text{wrist}}}\right) * a_{\text{wrist}} \tag{2}$$

where fw is the frequency weighting of ISO 5349-1 W_h, $TR(A_{\text{handle}}/A_{\text{wrist}})$ is the inverse of the transfer function from the tool handle on to the wrist, and a_{wrist} is the vibration magnitude on the wrist, including all affecting factors in Annex D of ISO 5349-1.

4. Experiment and Results and Discussion

The experiment was conducted using the same methods as paper [3], and the same experiment was used as in paper [2]. The experimental setup, procedure, and the subjects were the same as in these papers. The following, Table 1, shows the results.

Table 1. Test results summary (average of all participants).

Subject	Posture 1			Posture 2			Posture 3		
	On Tool	On Subject	TTS (dB)	On Tool	On Subject	TTS (dB)	On Tool	On Subject	TTS (dB)
Mean	5.33	8.82	20.42	3.95	8.51	17.50	3.44	10.57	21.88
SD	0.20	2.25	2.70	0.32	3.99	3.00	0.73	5.16	3.20

From the results of Table 1, it is believed that the TTS value increases with the vibration value of the tool handle vibration magnitude, according to ISO 5349-1, increases. As can be observed in Figure 1a,b of Paper [2], although the vibration value is small, the TTS value is large in the case of posture 3. From this, it is concluded that the vibration value evaluation method of the tool handle cannot be used to evaluate the effects on the human body. In addition, it is thought that the TTS value will increase according to the different posture, as shown in Table 1, so the value measured by the concept device [3], which is a concept that easily predicts the vibration value of the tool handle, corresponds to the TTS of the VPT. Therefore, it was clear that it can prevent HAVS in workers during tool work from the vibration value on the wrist, based on this paper's methodology [3].

Although the characterization of the vibration exposure currently uses the acceleration of the surface in contact with the hand as the primary quantity, it is reasonable to assume that the biological effects depend to a large extent on the coupling of the hand to the vibration source. It should also be noted that the coupling can considerably affect the measured vibration magnitudes. The vibration measurements should be made with forces which are representative of the coupling of the hand to the vibrating power tool handle or workpiece in typical operation of the tool or process.

Forces between the hand and gripping zone should be measured and reported. It is also recommended that a description of the operator's posture should be reported for individual conditions and/or operating procedures (see annexes D and F). The method for evaluation of vibration exposure described in this part of ISO 5349-1 [1] takes into account the vibration magnitude, the frequency content, the duration of exposure in a working day, and the cumulative exposure to data.

5. Conclusions

In the current study, the experiment was performed to clarify whether the a_{hv} can assess the risk from real tool work vibration exposure. From these experiments, the following conclusion was drawn:

Although the values from the test protocol and ISO 5349-1 a_{hv} values cannot apply to the all postures or subjects in real work conditions, it should be emphasized that the new evaluation method of the wearable equipment [3] can monitor the hand-transmitted vibration magnitude in real work for preventing HAVS, instead of the usage of the a_{hv} in the real work site.

Additionally, it was shown that it is necessary to carry out vibration measurement in a form that takes Annex D on-body vibration measurement equipment into consideration.

Author Contributions: Conceptualization, S.M., S.G. and Y.Y.; methodology, investigation, and original draft preparation, S.M.; review and editing, S.G. and Y.Y. All authors have read and agreed to the published version of the manuscript.

Funding: This research received no external funding.

Institutional Review Board Statement: The study was approved by the Edinburgh Napier University ethics committee.

Informed Consent Statement: Informed consent was obtained from all subjects involved in the study.

Data Availability Statement: The data are not publicly available due to privacy reasons.

Acknowledgments: We thank Mark Taylor (Edinburgh Napier University) for technical and instrumental support.

Conflicts of Interest: The authors declare no conflict of interest.

References

1. *ISO 5349-1*; Mechanical Vibration—Measurement and Evaluation of Human Exposure to Hand-Transmitted Vibration—Part 1: General Requirements. International Organization for Standardization: Geneva, Switzerland, 2001.
2. Taylor, M.; Maeda, S.; Miyashita, K. An Investigation of the Effects of Drill Operator Posture on Vibration Exposure and Temporary Threshold Shift of Vibrotactile Perception Threshold. *Vibration* **2021**, *4*, 395–405. [CrossRef]
3. Maeda, S.; Taylor, M.D.; Anderson, L.C.; McLaughlin, J. Determination of hand-transmitted vibration risk on the human. *Int. J. Ind. Ergon.* **2019**, *70*, 28–37. [CrossRef]
4. *ISO 8041-1: 2017*; Human Response to Vibration—Measuring Instrumentation—Part 1: General Purpose Vibration Meters. International Organization for Standardization: Geneva, Switzerland, 2017.
5. *ISO 8041-2: 2021*; Human Response to Vibration—Measuring Instrumentation—Part 2: Personal Vibration Exposure Meters. International Organization for Standardization: Geneva, Switzerland, 2021.

Proceeding Paper

French Occupational Disease System: Examples of Diseases Caused by Hand–Arm Vibration †

Anne Delépine

Department of Etudes et Assistance Médicales, Institut national de recherche et de sécurité (INRS), 75011 Paris, France; anne.delepine@inrs.fr; Tel.: +33-140-44-30-29

† Presented at the 15th International Conference on Hand-Arm Vibration, Nancy, France, 6–9 June 2023.

Abstract: In France on 25 October 1919, a law was created so that workers suffering from occupational diseases because of their work could receive specific compensation. Built on the notion of "presumption of origin", the occupational disease tables outline the conditions for recognition.

Keywords: occupational diseases; regulation; occupational diseases tables; procedure

Citation: Delépine, A. French Occupational Disease System: Examples of Diseases Caused by Hand–Arm Vibration. *Proceedings* **2023**, *86*, 28. https://doi.org/10.3390/proceedings2023086028

Academic Editors: Christophe Noël and Jacques Chatillon

Published: 13 April 2023

1. Introduction

Occupational diseases, which have constantly been on the rise for over more than 20 years now, can have an impact on both a person's professional life and private life. In France, employers are responsible for their prevention, but all occupational safety and health practitioners can play a role in identifying and preventing them to keep workers affected by these diseases in employment.

2. Definition

The French notion of an occupational disease is, above all, a medico-legal concept, conferring the right to specific compensation, which is identical to that of occupational accidents. This includes: a waiver of co-payments for all care relating to the disease including functional rehabilitation; a higher sick pay than that which is paid to people with a non-occupational-related sickness; the possible payment of benefits or annuities depending on the health effects according to the permanent impairment rating evaluated by Social Security's medical advisor. In France, costs are borne by employers through occupational accident/occupational disease contributions paid to the occupational hazard branch of Social Security by companies under the general social security scheme (RG) and to the agricultural mutual fund by companies under the agricultural social security scheme (RA). For civil servants, the system is managed by each individual administration.

Self-employed French workers are not covered for occupational hazards unless they have taken out voluntary insurance under the general social security scheme.

3. The Presumption of Origin

In the French system, a disease is occupational if it is the direct consequence of a worker's exposure to a physical, chemical, or biological hazard or resulted from the conditions under which they perform their professional activity. However, it can be difficult to establish a direct causal link between work and a given disease. This is why, since the introduction of a law on 25 October 1919, a disease has been recognized as occupational if it appears in one of the tables in the French Social Security Code or to the Rural Code. For civil servants, the disease tables under the general scheme also apply.

These tables all follow the same model (Table 1):

- A **title** stating the **hazard** that is taken into account. This can be a chemical agent (for example "Occupational diseases following the inhalation of asbestos dust" RG 30 or

RA 47), a biological agent (for example "Diseases due to tubercle bacilli and certain atypical mycobacteria" RG 40 or RA 16), a physical agent (for example "Diseases caused by ionizing radiation" RG 6 or RA 20), or a "work environment" (for example "Diseases caused by vibrations and shock transmitted by certain machine tools, tools and objects and repeatedly hammering objects with the palm heel" RG 69 or RA 29);

- A **left column** enumerating the **diseases or symptoms** recognized as possibly being linked to the hazard mentioned in the title (for example, for table RG 69: arthritis of the elbow, osteonecrosis of the lunate, or angioedema of the hands). Certain tables also mention the conditions of the diagnosis, such as those in table RG 42 "Hearing disorder caused by noise", or specific additional exams, such as functional tests for objective evidence of Raynaud's phenomenon for angioedema in part A of table RG 69;
- A **right column** presenting a **restrictive or indicative list of work** that involves exposure to the hazard and is likely to cause the disease or disorder mentioned in the left column. For table RG 69, this list is restrictive, i.e., only workers performing the work mentioned (work and tasks and not jobs) can have the benefit of the presumption of origin;
- Lastly, a **center column** stating the **time limit for compensation**, i.e., the maximum period between the end of exposure to the hazard (regardless of the reason for this end of exposure) and the first medical diagnosis. For some tables, there is also a **minimum period of exposure** to the hazard (as for part C of table RG 69).

Table 1. Example of an occupational disease table (RG 69). Diseases and disorders caused by vibration and shock transmitted by certain machine tools, tools, and objects and by repeatedly hammering objects with the palm heel. Date of creation: decree on 15 July 1980; the most recent update: decree on 6 November 1995 (Adapted from Ref. [1]).

Designation of Diseases and Disorders	Time Between End of Exposure and First Diagnosis	Restrictive List of Work Likely to Cause These Diseases and Disorders
- A -		
Osteoarticular diseases confirmed by radiology exams: -Arthritis of the elbow involving radiological signs of osteophytosis; - Osteonecrosis of the lunate (Kienböck disease); - Osteonecrosis of the carpal scaphoid (Köhler disease). Angioedema of the hand, predominately the index and middle fingers, which may be accompanied by cramps in the hand and prolonged impaired sensitivity and confirmed by functional tests for an objective diagnosis of Raynaud's disease.	5 years 1 year 1 year 1 year	Work regularly involving exposure to vibration transmitted by: (a) Hand-held machine tools, in particular: - Percussive tools, such as jack hammers, chipping hammers, bush hammers, and rammers; - Rotary–percussive machines, such as rock drills, hammer drills, and impact wrenches; - Rotary machines, such as polishers, grinders, sawing machines, and brush cutters; - Alternative machines, such as sanders and jigsaws. (b) Hand-held machines associated with certain abovementioned machines, particularly in chiseling work; (c) Workpieces held in the hand while being processed, particularly during grinding and polishing work and work on swaging machines.
- B -		
Osteoarticular diseases confirmed by radiological exams: -Arthritis of the elbow, involving radiological signs of osteophytosis; - Osteonecrosis of the lunate (Kienböck disease); - Osteonecrosis of the carpal scaphoid (Köhler disease).	5 years 1 year 1 year	Work regularly involving exposure to shocks caused by manual use of percussive tools: - Hammering work, such as smithing, sheet metal working, boiler making, and leather working; -Earth moving and demolition works; -Use of caulking guns;- Use of nail guns and riveting hammers.

Table 1. *Cont.*

Designation of Diseases and Disorders	Time Between End of Exposure and First Diagnosis	Restrictive List of Work Likely to Cause These Diseases and Disorders
- C -		
Hypothenar hammer syndrome (HHS) causing Raynaud's phenomenon or finger ischemia confirmed by arteriography, providing objective evidence of an aneurysm or thrombosis of the ulnar artery or the superficial palmar arch.	1 year (subject to an exposure period of 5 years)	Work regularly involving exposure to repeat hammering of objects with the palm heel or involving exposure to shock transmitted to the hypothenar eminence by a percussive tool or a tool receiving impact.

The presumption of origin enables a worker, covered by a French social security body, meeting the conditions of an occupational disease table, to not have to prove the link between their disease and their work; rather, it is up to the employer in France or the French social security body to demonstrate that there is no link between the disease and the professional activity.

The French occupational disease tables are drafted based on social consensus according to scientific expertise, within advisory commissions under the Labor and Agriculture ministries. They then become the subjects of simple decrees, applicable after being published in the Journal Officiel de la République française (official journal of the French Republic). They are created and updated based on scientific and technical developments. As of February 2023, there were 118 tables for the general social security scheme and 61 for the agricultural scheme.

4. French Complementary Health Insurance

In certain instances, despite non-compliance with conditions mentioned in the tables, the employee's disease can still be recognized as having an occupational origin, either because the criteria in the center and right columns are not met if a direct link is established with the professional activity (paragraph 6 in Article L. 461-1 of the Social Security Code [2]), or because the disease does not currently appear in a table and has caused the death of the victim or an impairment rating of at least 25% and a direct and essential link is established between this disease and the work (paragraph 7 in Article L. 461-1 of the Social Security Code [2]). The establishment of these links falls within the jurisdiction of the regional occupational disease recognition committee.

5. Procedure

First, it is necessary to establish a precise diagnosis. To achieve this, in view of the symptoms suggestive of a disease, an occupational physician generally addresses the worker to a medical specialist who will conduct exams to confirm and specify the diagnosis.

In all cases (presumption of origin or complementary system), it is up to the victim or their beneficiaries to make the request for recognition with their French social security body or administration where civil servants are concerned. Examination of the request must follow the adversarial principle (employer/worker).

6. The Occupational Diseases Most Frequently Recognized in France

For more than 20 years now, from about 80% to 85% of occupational diseases recognized by social security bodies are those affecting the musculoskeletal system (musculoskeletal disorders (MSDs)) and corresponding to tables RG 57 or RA 39 "Peri-articular diseases caused by certain work movements and postures", RG 69 or RA 29 "Diseases caused by vibration and shocks transmitted by certain machine tools, tools and objects and by repeatedly hammering objects with the palm heel", RG 79 or RA 53 "Chronic injury of the meniscus", RG 97 or RA 57 "Chronic disorders of the lumbar spine caused by low- and medium-frequency vibration transmitted to the entire body", and RG 98 or RA 57bis "Chronic disorders of the lumbar spine caused by regular manual handling of heavy loads".

Between 2016 and 2021, there were an average 96 occupational diseases recognized per year under table RG 69. Osteonecrosis of the lunate (34%) and elbow arthritis (33%) are the most frequent disorders. Vascular disease represents 20% and angioedema roughly 10% of the disorders. Osteonecrosis of the scaphoid is anecdotal (slightly less than 3%). The sectors most frequently concerned are construction (structural work and finishings), civil engineering, and the automobile industry (manufacturing and repairs).

For diseases that do not appear in the tables, certain mental health effects (severe depression, generalized anxiety, and post-traumatic syndrome) resulting from psychosocial risks are at the top of the list, with several hundreds of cases recognized each year, which have been significantly more common in the past few years.

All of the tables, together with comments, can be consulted on www.inrs.fr/mp (accessed on 11 April 2023).

Funding: This research received no external funding.

Institutional Review Board Statement: Not applicable.

Informed Consent Statement: Not applicable.

Data Availability Statement: Not applicable.

Conflicts of Interest: The author declares no conflict of interest.

References

1. Affections Caused by Vibration and Shock Transmitted by Certain Machine Tools, Tools and Objects and by Iterative Shocks of the Heel of the Hand on Fixed Parts—Table Equivalent 29 in Agricultural Regime. Available online: https://www.inrs.fr/publications/bdd/mp/tableau?refINRS=RG%2069 (accessed on 11 April 2023). (In French)
2. *Article L. 461-1-Social Security Code*; French Government: Paris, France, 2018. (In French)

Proceeding Paper

Daily Exposure to Hand-Arm Vibration of Technicians in Wastewater Treatment Plants and After-Sales Service [†]

Rémi Petitfour *[iD], Guenaëlle Ducrot and Isabelle Jannin Devilleneuve

Association interprofessionnelle des Centres Médicaux et Sociaux de santé au travail de la région Ile-de-France (ACMS), 92158 Suresnes, France; guenaelle.ducrot@acms.asso.fr (G.D.); isabelle.jannin-devilleneuve@acms.asso.fr (I.J.D.)

* Correspondence: remi.petitfour@acms.asso.fr

† Presented at the 15th International Conference on Hand-Arm Vibration, Nancy, France, 6–9 June 2023.

Abstract: This study encompasses two implementations of evaluation of the daily exposure to hand-arm vibration of employees. The first one focuses on maintenance technicians and operating agents in wastewater treatment plants and as network technicians on the water distribution network. The second one focuses on technicians in the after-sales service of a company supplying a wide range of percussive and vibrating portable tools for professionals and the general public. In both cases, we established a precise inventory of the tools used by the employees, and computed the duration of use per day of each tool. We used the INRS tools N43 "Calculette Vibrations Mains Bras" and N59 "OSEV—Vibrations transmises aux membres supérieurs" to determine the daily exposure A(8). These results were explained to the employees, with practical advice for prevention, collectively during meetings and individually during sensitization workshops.

Keywords: daily exposure assessment; hand-arm vibration; maintenance technician; wastewater treatment plant; vibrating hand tools; percussive hand tools; sensitization workshop

1. Introduction

We present the first implementation of the assessment of hand-arm vibration exposure of maintenance and operations technicians working in wastewater treatment plants and network technicians working in the water distribution network; all of these workers use daily percussive and vibrating portable tools. The results of the evaluation of their daily exposure were delivered during a sensitization workshop.

We then present a second implementation that concerns the exposure to hand-arm vibration of technicians in the after-sales service of a company supplying a wide range of portable percussive and vibrating tools for professionals and the general public, who diagnose, repair, test and return the equipment received from customers. The results were presented at a meeting with the quality department, the employees concerned and members of the Economic and Social Council (CSE).

2. Method

In both implementations, we establish a precise inventory of the tools used by the employees, completed with the duration of use per day of each tool by questioning the employees and the team leaders.

We use a "Hand Arm Vibration Calculator" [1] to compute the duration of exposure to reach the action value (2.5 m/s^2) and the duration of exposure to reach the limit value (5 m/s^2), using the vibration coefficient (or vibration level) given by the supplier in its technical data sheet and the estimated duration of exposure per day. We elaborated a table to deliver all the results; an extraction is shown in Table 1.

Citation: Petitfour, R.; Ducrot, G.; Devilleneuve, I.J. Daily Exposure to Hand-Arm Vibration of Technicians in Wastewater Treatment Plants and After-Sales Service. *Proceedings* **2023**, *86*, 38. https://doi.org/10.3390/proceedings2023086038

Academic Editors: Christophe Noël and Jacques Chatillon

Published: 21 April 2023

Table 1. Results for a vibrative tool and a percussive tool extracted from the table.

	Designation of Portable Tool	Vibration Level (m/s^2)	Duration of One Test of the Portable Tool (sec)	Activity Area	Percussive or Vibrating Tool	Hand-Arm Daily Vibration Exposure A(8) m/s^2 (for 1 h of Exposure)	Duration of Vibration Exposure Pour Action Value (2.5 m/s^2)	Duration of Vibration Exposure Pour Action Value (5 m/s^2)	Number of Tests for Limit Value (5 m/s^2)
	Circular saw	1.5	20	Metal ou wood cutting	Vibrative	0.5	>8 h	>8 h	>100
	Rotary hammer	12.5	120	Construction works	Percussive	4.4	0 h 19	1 h 17	39

We computed the daily exposure A(8) from the information provided by the employees utilizing the devices, including their estimated duration of exposure per day, using a simplified tool for evaluation of exposure of vibration transmitted to the upper limbs [2].

The results of the hand-arm OSEV tool were placed in a report intended for the quality manager of the company. As an example, Figure 1 displays the results for the angle grinder tool.

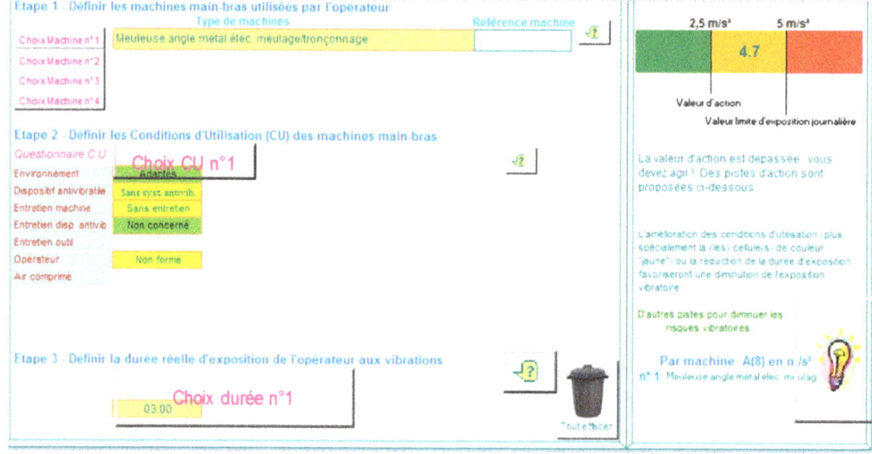

Figure 1. Result of the hand-arm OSEV tool for an angle grinder tool (meuleuse d'angle).

3. Results

For technicians working in wastewater treatment plants and network technicians working in the water distribution network, we assessed the eight most used tools with the OSEV tool. For seven of them, the daily exposure A(8) was between the action value (2.5 m/s^2) and limit value (5 m/s^2).

For technicians in the after-sales service of a company supplying a wide range of portable percussive and vibrating tools, we inventoried 262 portable tools in the table and filled it with the results of the "Hand Arm Vibration Calculator" for each of these tools. For 200 out of 262 (76.3%), the daily exposure A(8) was below the action value (2.5 m/s^2). For 46 of them (17.6%), the daily exposure A(8) was between the action value (2.5 m/s^2) and limit value (5 m/s^2). For 16 of them (6.1%), the daily exposure A(8) was above the limit value (5 m/s^2).

We could highlight the tools with exceedances of action values and limit values and give prevention advice to users to lower their exposure level, collectively during meetings and individually during an awareness workshop. Recommendations were based on flyer

"Syndrome des vibrations La main et le bras en danger" [3] (10/2016) and document "Bien choisir son outil portatif pour mieux travailler" [4] (03/2017).

Author Contributions: All authors contributed equally. All authors have read and agreed to the published version of the manuscript.

Funding: This research received no external funding.

Institutional Review Board Statement: Not applicable.

Informed Consent Statement: Not applicable.

Data Availability Statement: The data presented in this study are available on request from the corresponding author.

Conflicts of Interest: The authors declare no conflict of interest.

References

1. INRS Tool N43 Hand Arm Vibration Calculator. Available online: https://www.inrs.fr/risques/vibration-membres-superieurs/evaluation-risque.html (accessed on 3 April 2023).
2. INRS Tool N59 OSEV—Vibrations Transmitted to the Upper Limbs. Available online: https://www.inrs.fr/media.html?refINRS=outil59 (accessed on 3 April 2023).
3. INRS Flyer ED 6204 Syndrome des Vibrations La Main et le Bras en Danger. Available online: https://www.inrs.fr/media.html?refINRS=ED%206204 (accessed on 3 April 2023).
4. CRAMIF Brochure DTE 271 Bien Choisir Son Outil Portatif Pour Mieux Travailler. Available online: https://www.cramif.fr/actualites/bien-choisir-son-outil-portatif-pour-mieux-travailler (accessed on 3 April 2023).

Proceeding Paper

Zero Vibration Injuries—A Swedish Holistic Approach to Reduce Vibration Injury [†]

Carolina Pettersson *, Hans Lindell and Snævar Leó Grétarsson

RISE Research Institutes of Sweden, 431 53 Mölndal, Sweden; hans.lindell@ri.se (H.L.);
snaevar.gretarsson@ri.se (S.L.G.)
* Correspondence: carolina.pettersson@ri.se
† Presented at the 15th International Conference on Hand-Arm Vibration, Nancy, France, 6–9 June 2023.

Abstract: Vibration injuries cause significant costs for society, great personal suffering, and often the relocation of personnel within a company. The project "Zero Vibration Injuries" is a Swedish initiative with the objective of taking a holistic approach to the problem, involving all stakeholders. The project's vision is "Zero Vibration Injuries". This is achieved by addressing the source of the problem by reducing the vibration levels in hand-held machines and applying the solutions in industry to the benefit of the users.

Keywords: vibration injuries; hand-held machines; preventive measures; vibration reduction

1. Introduction and Background

Every day, more than 400,000 people in Sweden are exposed to vibrating machines for at least two hours per day. Injuries due to the effects of vibrations were the most common occupational injury for men during 2016–2020 [1]. If carpal tunnel syndrome, which often is related to vibration injuries, is included in this, the percentage of occupational injuries linked to vibrations is almost 60 percent among men and almost 20 percent among women. Currently, several industry sectors do not comply with EU directives [2,3] for vibration exposure, often due to no availability of machines with low vibration levels. Even in areas where there are effective technical solutions, the lack of knowledge is an important factor leading to constant or increasing numbers of vibration injuries.

Since the work within the project Zero Vibration Injuries started in 2014, the purpose has been to reduce vibration injuries by addressing the source of the problem and developing low-vibration machines. The objective is to take a holistic approach to the problem, with all stakeholders in society participating in the project, i.e., machine manufacturers, a comprehensive range of machine users, the Swedish work environment authority, employer and labor organizations, and occupational medicine and vibration researchers.

The strategy has been to develop low vibration concept prototypes, representing the main research area. It has been shown for a broad range of types of machines that machines do not need to vibrate and thus injure people, and this is also our motto within the project. At this stage of the project, lab-scale prototypes and concepts are being scaled up to establish replacements for new and existing hand-held machines, with the goal to make them accessible on the market.

In this project, vibration regulated with ISO 5349-1 with an upper frequency limit of 1250 Hz has been addressed, but also higher frequency vibrations, so-called ultravibrations, have been reduced. Ultravibrations are vibrations with a frequency higher than 1250 Hz, which is beyond the human perception threshold. Precautions against high frequency shock vibration are important, since there is a substantial risk for vibration injuries [4–7]. Reducing ultravibrations is, from a technical perspective, generally much easier than those at lower frequencies due to the laws of physics. Regrettably, currently there is no incentive for machine manufacturers to reduce these vibrations since there is no standardized method to measure and quantify ultravibrations. However, work from the International Organization for Standardization, ISO, is ongoing in order to include high frequencies and transients/shocks.

2. Method

The project has, since the start 2014, been divided into three stages, and these are:

1. It can be done!Representative machine types have been redesigned and tested in the laboratory.
2. It can be done in real production!The solutions from stage one have been upscaled and set into industrial production in the relevant areas.
3. Make it happen!The developed solutions will be implemented for the project participants, leading to the manufacturing of low-vibration machines as requested by the users, which results in lower vibration exposure and injuries.

In the current stage 3, the work is being conducted in real industrial environments with participating parties representing the automotive, construction, stone, and dental sectors.

2.1. Activities

The work is based on three main activities:

- Information on low-vibration alternatives;
- Facilitating the selection of low-vibration machines;
- Establishing a culture change.

Information on low-vibration alternatives: We have shown that vibrations can be reduced with new techniques on the lab scale and in prototypes. Now, scaling up the vibration-reducing solutions to system solutions for new and existing hand-held machines and machine groups is the next step. One example of doing this is through the newly formed company, ATVA License Group, which has been established for commercializing the ATVA technology worldwide (see more about ATVA below in Section 2.2).

Facilitating the selection of low-vibration machines: In order to enable machine users to require, assess, and purchase low-vibration machines, measurement and assessment of vibrations, including the peak vibrations in the high frequency area, are needed. Ultravibrations are not measured in vibration instruments used in the field today. However, in this project, RVM10, a measurement system for hand-arm vibrations capable of simultaneously measuring vibrations according to ISO 5349-1 and ultravibrations up to 30 kHz is being developed. The system is a scientific prototype, with the goal to make it available for purchase.

Establishing a culture change: This concerns company behavior regarding vibrations and vibration injuries, also including ultravibrations, and minimizing the problem by choosing low-vibration machines. Knowledge and information dissemination of vibration-reducing solutions, understanding and acceptance throughout the organization of the vibration problem, and raising awareness about the risks of injury from ultravibrations are important in all levels of organizations, e.g., machine users, managers, purchasing departments, occupational health care, labor organizations, employers, and so on.

2.2. Methods to Reduce Vibrations in Hand-Held Machines and Tools

Vibrations are reduced in machines by design solutions based on (a) balancing rings, also called autobalancers, for vibrations originating from imbalance and rotational forces in, e.g., grinders, (b) Auto Tuning Vibration Absorbers (ATVA), which reduce reciprocating forces in, e.g., rockdrills, and (c) traditional vibration isolation and conceptual redesign. These solutions, alone or in combination, can be used for the vast majority of hand-held vibrating machines.

The balancing ring is in principle a ring containing balls and a dampening lubricant. The balls will automatically adjust themselves and compensate for the imbalance in a fraction of a second and keep the system continuously in balance. As soon as new imbalances are created, the balls immediately readjust [8].

The Auto Tuning Vibration Absorber (ATVA) technology [9–11] is a vibration absorbing unit that is integrated into a machine with reciprocating vibrations. The ATVA unit creates a counter force to the excitation force from the piston of the machine, which reduces the vibrations.

Traditional techniques, e.g., vibration isolation of handles and conceptual redesign, are also used to reduce vibrations in machines and tools when applicable.

3. Results

Several machines and tools have been re-built within the project, according to the methods mentioned above (Section 2.2). A few examples, which represent a vast proportion of machines that cause vibration injuries, are given in Table 1 below. The project also includes dental tools.

Table 1. Summary of some machines and tools, the reduction methods, and the consequent reduction in vibration levels.

Machine	Reduction Method	Vibrations Before (m/s^2)	Vibrations After (m/s^2)	Reduction
Round vibratory plate	Optimized vibration isolation	11.1 [1]	5.5 [1]	51%
Chisel machine	ATVA, optimized vibration isolation	20 [1]	2.7 [1]	87%
Rock drilling machine	Isolating handle	27.3 [1]	8.6 [1]	68%
Rammer	ATVA, isolating handles	25–32 [1]	5–8 [1]	75–80%
Anvil	Vibration isolation	13 [1] 8000 [2]	6 [1] 150 [2]	54% 98%

[1] Vibrations measured according to ISO 5349-1; [2] Ultravibration, VPM, average peak measured up to 30 kHz.

Peak ultravibrations were measured up to 30 kHz. An average level was estimated by visual assessment; however, it is recommended that peak ultravibrations should be calculated according to the VPM algorithm [12].

4. Discussion and Conclusions

It has been shown that machines and tools can be re-built or re-designed in order to minimize harmful vibrations, often at a low cost and potentially enabling additional improvements at the same time. It is important to spread the knowledge about the problem and the solutions in order to reduce vibration-related injuries.

Some of the re-built machines and tools are in serial production. The newly developed, low-vibration machines meet the specific requirements of users. Demand from workers and companies are necessary for the manufacturers to start producing more low-vibration machines. The prerequisite is standardized methods for vibration measurements, which also includes ultravibrations and shocks. A leap forward could come from the suggested revised European Machinery Regulations, planned to come into force from Q1 2023, requiring manufacturers to declare these vibrations. This will hopefully lead to an increased drive from the machine user to demand low-vibration machines and an incentive for the machine producers to reduce them. It is necessary that low vibration machines are demanded, manufactured, and purchased. Machines do not need to vibrate and injure people!

Author Contributions: Writing—original draft preparation, C.P. and H.L.; project administration, C.P.; methodology, S.L.G. and H.L.; funding acquisition, H.L. All authors have read and agreed to the published version of the manuscript.

Funding: This research was funded by the Swedish Innovation Agency Vinnova, grant number 2021-01859.

Institutional Review Board Statement: Not applicable.

Informed Consent Statement: Not applicable.

Data Availability Statement: Not applicable.

Conflicts of Interest: The authors declare no conflict of interest.

References

1. *Allvarliga Arbetsskador och Långvarig Sjukfrånvaro*; AFA Försäkring: Stockholm, Sweden, 2022. (In Swedish)
2. EU Machine Directive, 1989/392/EC. Available online: https://eur-lex.europa.eu/legal-content/EN/TXT/?uri=CELEX%3A32006L0042 (accessed on 2 April 2023).
3. EU Vibration Directive, 2002/44/CE. Available online: https://eur-lex.europa.eu/EN/legal-content/summary/exposure-to-mechanical-vibration.html (accessed on 2 April 2023).
4. Barregard, L.; Ehrenström, L.; Marcus, K. Hand-arm vibration syndrome in Swedish car mechanics. *Occup. Environ. Med.* **2003**, *60*, 287–294. [CrossRef]
5. Lindell, H.; Lönnroth, I.; Ottertun, H. Transient Vibration from Impact Wrenches: Vibration Negative Effect on Blood Cells and Standards for Measurement. In Proceedings of the Eighth International Conference on Hand-Arm Vibration, Umeå, Sweden, 9–12 June 1998.
6. Ando, H.; Nieminen, K.; Toppila, E.; Starck, J.; Ishitake, T. Effect of impulse vibration on red blood cells in vitro. *Scand. J. Work Environ. Health* **2005**, *31*, 286–290. [CrossRef] [PubMed]
7. Govinda Raju, S.; Rogness, O.; Persson, M.; Bain, J.; Riley, D. Vibration from a Riveting Hammer Causes Severe Nerve Damage in the Rat Tail Model. *Muscle Nerve* **2011**, *44*, 795–804. [CrossRef] [PubMed]
8. Lindell, H. Vibration reduction on hand-held grinders by automatic balancing. *Cent. Eur. J. Public Health* **1996**, *1*, 43–45.
9. Lindell, H.; Berbyuk, V.; Josefsson, M.; Grétarsson, S.L. Nonlinear Dynamic Absorber to Reduce Vibration in Hand-held Impact Machines. In Proceedings of the International Conference on Engineering Vibration, Ljubljana, Slovenia, 7–10 September 2015.
10. Lindell, H. Attenuation of Hand-held Machine Vibrations, Application of Non-linear Tuned Vibration Absorbers. Thesis for the Degree of Licentiate of Engineering, Chalmers University of Technology, Gothenburg, Sweden, 2017.
11. Lindell, H. Impact Machine. Patent WO 2014/095936 A1, 21 October 2015.
12. Johannisson, P.; Lindell, H. Definition and Quantification of Shock/Impact/Transient Vibrations. *arXiv* **2022**, arXiv:2211.08999.

MDPI
St. Alban-Anlage 66
4052 Basel
Switzerland
Tel. +41 61 683 77 34
Fax +41 61 302 89 18
www.mdpi.com

Proceedings Editorial Office
E-mail: proceedings@mdpi.com
www.mdpi.com/journal/proceedings